L.E.O. BRAACK

FIELD GUIDE TO INSECTS OF THE KRUGER NATIONAL PARK

This book has been produced by Struik Publishers, Cape Town,
in collaboration with the National Parks Board, Pretoria

Cover photographs: (top left) handmaiden moth (Family Ctenuchidae) (L.E.O. Braack); (top right) dung beetle (Family Scarabaeidae) (L. Hoffmann); (bottom left) caterpillar of hawk moth (Family Sphingidae) (L. Hoffmann); (bottom right) praying mantids (Order Mantodea) (L. Hoffmann).

Struik Publishers
(a member of the Struik Group (Pty) Ltd)
Struik House
Oswald Pirow Street
Foreshore
Cape Town
8001

Reg. no.: 63/00203/07

First published 1991

Copyright © text: L.E.O. Braack 1991
Copyright © photographs: L.E.O. Braack and the respective photographers listed below 1991.
Copyright © vegetation map: Euan Waugh 1991

Illustrations on page 65 (del. T.J.D. Coates) reproduced from 'Status of the Taxonomy of the Hexapoda of Southern Africa.' (ed. W.G.H. Coaton) *Entomology Memoir.* No 38., under Government Printers Copyright Authority 9188 of 25 February 1991.
Anatomical illustration on page 149 by Paula Chasty reproduced from *Insects of the World* (1984) by A. Wootton, (Blandford, a division of Cassell, plc), with kind permission of the Publisher.

Edited by Tracey Hawthorne
Colour plates designed by Janine Poezyn
DTP conversion by BellSet, Cape Town

Reproduction by Fotoplate (Pty) Ltd, Cape Town
Printing and binding by National Book Printers, Goodwood

All rights reserved. No part of this publication may be reproduced, stored in a retrieval system or transmitted, in any form or by any means, electronic, mechanical, photocopying, recording or otherwise, without the written permission of the copyright owners.

ISBN 0 86977 945 1

PHOTOGRAPHIC CREDITS
With the following exceptions the photographs for this book were supplied by the author, L.E.O. Braack:
A. Bannister (ABPL) (Termitidae) male and female termites Pl. 3; (Cicadellidae) leafhopper Pl. 7; (Aphididae) aphids Pl. 8; (Coccidae) soft scale insects Pl. 9; (Pentatomidae) shield bug Pl. 10; (Cicindelidae) tiger beetle Pl. 15; (Gyrinidae) whirligig beetle Pl. 15; (Tabanidae) horse fly Pl. 25; (Chalcidoidea) chalcid wasp Pl. 39 and (Eumenidae) potter wasp Pl. 40.
L. Hoffmann dung beetle (cover); caterpillar of hawk moth (cover); and praying mantids (cover).
A. Weaving silverfish Pl. 1; (Ascalaphidae) owl fly Pl. 14; (Charaxidae) *Charaxes brutus* Pl. 37 and (Danaidae) African monarch butterfly Pl. 38.
Copyright for all photographs reproduced in this book remains with the owners.

Contents

Foreword 4
Acknowledgements 5
How to use this book 6

The place of insects in the animal kingdom 7
The importance of insects to man 8
   Insects of medical importance 8
   Insects in agriculture 8
   Beneficial insects 9
Insects and the Kruger National Park 10
Key to Afrotropical Orders of Hexapods 11
Vegetation map of the Kruger National Park 21
Colour plates 22-64

Classes Protura, Diplura and Collembola 65
Class Insecta 65
   Archaeognatha and Thysanura 65
   Ephemeroptera 66
   Odonata 66
   Phasmatodea 68
   Mantodea 68
   Blattodea 69
   Isoptera 70
   Zoraptera 73
   Orthoptera 73
   Dermaptera 77
   Embiidina 78
   Plecoptera 78
   Psocoptera 79
   Phthiraptera 79
   Hemiptera 82
   Thysanoptera 90
   Megaloptera 91
   Neuroptera 91
   Siphonaptera 93
   Coleoptera 95
   Strepsiptera 108
   Mecoptera 108
   Diptera 109
   Trichoptera 123
   Lepidoptera 124
   Hymenoptera 135

Glossary 146
Bibliography 150
Index to common names 152
Index to scientific names 154
Index to Afrikaans names 158

Foreword

Because we are attracted to the dominant size and majesty of larger animals such as lions and elephants, visitors to the Kruger National Park tend to judge the success of their trip by how many of the 'big five' they encounter – and yet these magnificent beasts survive in part due to the myriad insects which effectively maintain most ecosystems on earth.

Insects are the sole agents for pollinating a wide variety of trees and other plants, mopping up at carcasses, feeding on dung and decaying vegetation, preying on and parasitizing themselves and other organisms to maintain a healthy balance of numbers, and making themselves available as food to fish, frogs, reptiles, birds and even mammals. They are the organisms that cause the death or disability of millions of people annually by transmitting malaria and other diseases and bring about crop losses worldwide of up to 30 per cent, but also have beneficial effects without which it is probable that all terrestrial tropical and temperate ecosystems would collapse.

In recent years there has been an increasing awareness and appreciation by the public of not only the importance of the smaller denizens of the bush but also the beauty and fascinating life histories of many insects. The graceful poise and spectacular colouring of a butterfly sipping nectar on a flower can rival those of any bird, and the ferocity and hunting skills of the tiger beetle are, albeit on a smaller scale, as formidable as those of its mammalian namesake.

The Kruger National Park is blessed with a great abundance of insects, and these are easily studied while they go about their business on flowers and trees, around pools and on the bare ground in camps and at picnic sites. A quick search alongside the road near a flowering shrub or vegetation-lined pan will often reveal a whole world of insects busily going about their lives, and provide an opportunity to view a wide variety of different insect families. Trailists walking in the wilderness areas are particularly well positioned to experience all the facets of insect life.

I have no hesitation in recommending this well illustrated book, which deals with all the commonly encountered groups of insects visitors to the Park are likely to see. It is, of course, impossible in a book of this size to illustrate each species, as this would require several thousand photographs, but the representative illustration for each family or group of insects should enable the visitor to locate most insects, and to learn about their general biology and habits. This book is a valuable addition to our series of guides to the wildlife of the Kruger National Park, which serves to increase the benefit and enjoyment our visitors derive from their contact with our wildlife heritage.

DR U DE V PIENAAR
CHIEF DIRECTOR: NATIONAL PARKS BOARD

Acknowledgements

For those wonderful weekends spent collecting insects at Nwanetsi and elsewhere while I was still floundering in primary school, I am very grateful to my brother Harold and his wife Toni. To my parents who gave me the encouragement, the means and the freedom to spend countless hours traipsing around with a butterfly net, as always a special thank you. My gratitude also to professors Ted Bosman and Denis Brothers and doctors Bill Overal and Ray Miller, who provided me with the foundations of entomology and have become friends. Denis Brothers also contributed his expertise by compiling the key to the various insect orders, and he and Ray Miller kindly read the manuscript. To Dr Valerius de Vos, thank you for the rare opportunity to work in the Kruger National Park, and for your constant readiness to help. To the staff of the Kruger National Park's Department of Nature Conservation, you have showed and taught me wondrous things: thanks very much. To Merle Whyte, and sometimes Lollie de Jager and Tina van Niekerk, who typed the manuscript and put up with my numerous scribbled additions and alterations, you did a marvellous job. Also a very warm thank you to my companions during many years of fieldwork, Philemon and Alfred Nkuna, Solomon Monareng, and the others: we shared many good times and unusual situations.

I owe a special word of gratitude to Dr U de V Pienaar, Chief Director: National Parks Board, for granting me the opportunity to write this book, and also for the encouragement and kind words during my schooldays when butterfly collecting was a shared interest. This book was his idea and is a tribute to his wide-ranging interests and genuine concern for all forms of wildlife. Piet van Wyk deserves thanks, as does the National Parks Board, for helping to make this book reality. Finally, thanks to the staff of Struik for their professionalism, and Professor Clarke Scholtz for his assistance with the Afrikaans terms in the glossary.

LEO BRAACK
SKUKUZA, FEBRUARY 1991

How to use this book

I think most people will agree that the more one learns about a particular subject the more interesting that subject becomes. This is one of the principal aims I hope to achieve through this book: that more of us will come to enjoy the beauty inherent in many dragonflies, beetles, butterflies and other insects, and the fascinating ways in which most insects go about their daily lives. This is an interest people can pursue not only in a game reserve, but in their own back yard.

It would be impossible to describe and illustrate each of the literally thousands of insect species which occur in the Kruger National Park, especially as not all are known to science and numerous new species have yet to be named. Since the general appearance and biology of the insects in a particular group tend to be similar, I have chosen a middle path and discussed the broad unifying features of orders and families; a photograph of an insect which is fairly representative of each group is used as an illustration. There are certain insects – processionary worms, for example – which are particularly striking or interesting and thus merit discussion at species level. Here I hope I have provided sufficient detail to satisfy the curiosity which many of them evoke.

While most families of insects are readily distinguishable by general appearance, habits, or even an intuitive combination of physical attributes and lifestyles, sometimes the differences are not so obvious. For this reason where very close similarities in appearance and habits exist in two families I may discuss these families together, such as the neuropteran families Hemerobiidae and Chrysopidae; treating them this way makes for easier comparison. Not only is this the case with families but it may sometimes be so even with entire orders, such as Thysanura and Archaeognatha.

If you wish to find the relevant discussion on a particular insect, you can take one of two routes. The more arduous one – and one which generally requires a dead specimen, sometimes a microscope, and a certain familiarity with insect anatomy – is to use the Key to Afrotropical Orders of Hexapods on page 11. A scientific key is a branching sequence of decisions, presented as alternatives. Those who are unfamiliar with their use may find the following simple example helpful:
1. With wings 2
   Without wings 3

Does the insect have wings? If so, then go to number two; if not then go to number 3. The key is read in this way until the name of an order is reached. (The bracketed numbers encountered in the key indicate in each case the number of the foregoing chosen couplet.) This should enable a reliable identification at least to the level of insect order, but is intended more for laboratory use by undergraduate students of entomology.

The easier route – and one employed in most wildlife guidebooks – is simply to go through the illustrations until you find one which most closely resembles the insect you are curious about, and then note its name and the page number on which it is described. This works in most cases, but be warned: because of the great diversity and range of adaptations often present even within a single family, you might not always be able to classify your insect. Regard this as a challenge. Despite the occasional frustrations and disappointments, the pleasure gained from learning more about these interesting creatures is well worth the effort. As you become more familiar with family and other names, you might find it useful to refer to the Index on page 152.

Often in the text I refer to the insects in a particular group as being 'small', 'moderate' or 'large' in size. While these terms might seem vague, it would be most presumptuous of me to attempt to provide cut-and-dried, finite measurements encompassing all species in each group. Firstly, there is probably not a single family of insects in which new species do not await discovery, and these species might well fall outside the range indicated for the family, so imposing size limits may be misleading; and secondly, insects often vary in size depending on how much food they had earlier in life: adult blowflies which had ample resources in the maggot stage, for example, might be twice the size of those adults which were starved as immatures. Despite this, however, many insect groups do have an approximate average size for species in the group. As a rough guide, 'small' insects are less than 3 mm in length while 'large' insects are those exceeding 15 mm.

Interspersed throughout the text are occasional technical terms which will probably be unfamiliar to many people. I have kept these to a minimum, and the meanings of those I have used are defined in the Glossary on page 146.

Similarly, and again these have been kept to a minimum, I occasionally refer to other authors in the text, and these references are usually followed by a date in brackets, eg. 'Southwood (1977)'. This is generally to indicate the source of a particular snippet of information and the year in which it was published. Full details of the relevant book or journal are provided under that author's name in the Bibliography on page 150. The Bibliography also serves as an introduction to some of the articles published on insects of the KNP, but I must emphasize – or risk alienating the many fellow entomologists and friends who have published technical work done on insects in the KNP – that this is by no means an exhaustive list, and I hope those entomologists will forgive any omissions.

Some people might find it strange that I have included certain orders (eg. Zoraptera and Megaloptera) or other groups of insects in this book despite mentioning that they do not occur in the KNP. These inclusions are, firstly, to make readers aware of these groups since they may be discovered in the KNP at some future date, or because the reader should be aware of these major groups when using this book elsewhere (it will be equally useful anywhere else in southern Africa); and secondly, these orders are referred to in the Key to Afrotropical orders of hexapods, this key having been compiled to embrace all insects likely to be found anywhere in Africa south of the Sahara. Brief discussions of these groups therefore seemed justified.

Conversely, while all orders occurring in the KNP are discussed, not all the families within each order are mentioned. Many orders (eg. Coleoptera and Lepidoptera) contain numerous families and although some of these families may have representative species in the KNP, they are not discussed either because of their extreme rarity or their obscure nature. Adding these groups would have more than doubled the length of the book without really adding much information; only the more common, striking groups are therefore discussed.

Finally, the publishers have been considerate in including a vegetation and route map of the KNP, which visitors should find useful for orientation and perspective.

The place of insects in the animal kingdom

The animal kingdom is divisible into vertebrates, of which mammals, birds, reptiles and amphibians form a part, and invertebrates, containing animals such as snails, earthworms, roundworms, sponges, and the Arthropoda. The Arthropoda include all animals that have jointed limbs and a chitinized exoskeleton – a toughened outer skin. It is in this group that we find the classes Crustacea (shrimps and crabs), Myriapoda (centipedes and millipedes), Arachnida (spiders and scorpions), a few other, lesser known, classes – and the Insecta.

What, exactly, is an insect? Defined scientifically, insects are exognathous, hexapod Arthropoda. That is, their mouthparts are situated outside the buccal cavity or mouth, they are six-legged animals and they have jointed, membranous limbs. Insects are predominantly terrestrial, living in and on the soil, but they can also be found in the air, in water, in and on plants, living as parasites in the bodies of other animals or externally as bloodsuckers, or even at sea. If considered in terms of diversity (numbers of species) and abundance (numbers of individuals), insects are by far the most successful organisms on earth. There are more species of insects than all the other animal species put together – over 70 per cent of the species in the animal kingdom are insects. They are the only organisms that present a serious threat to humans' available food resources, forcing us to wage a constant battle against them to protect our agricultural crops and stored food products.

A great deal of the success of insects can be attributed to their power of flight. This enables them to more effectively escape from predators, to cover large areas in a relatively short time and so greatly enhance their opportunities of finding food and mates, and to rapidly discover newly opened or newly formed habitats such as a carcass or pool of water and so out-compete many potential competitors by early colonization of the habitat. Another factor contributing to their success, and the success of other arthropods, is their hard exoskeleton. This not only provides protection from the sun but also allows the legs to operate as a system of levers. (To understand this better, try to imagine how a soft-bodied organism like an earthworm would efficiently operate long legs. It simply would not be able to.)

The importance of insects to man
Insects of medical importance

In man's recent history, the greatest single mass death of humans occurred as a result of the bubonic plague of 1348–9, sometimes also referred to as the black death, the epidemic that killed almost half of the entire population of England. The disease organism which caused this near genocide is a bacterium, *Yersinia pestis*, endemic in rat populations on several continents. Rat fleas, especially those of the genus *Xenopsylla*, carry the bacteria from one rat to another when they change hosts. Under dirty and crowded conditions, as often existed during the Middle Ages, these rats came into close contact with humans. The fleas then jumped on to people and, when feeding, regurgitated the bacteria into the human bloodstream. In modern times people have adopted more hygienic living habits and today have the added protection of insecticides and rodenticides to kill off both fleas and rats.

Another major killer of man, and a far more familiar one, is malaria. In the case of this disease the host is man, the disease organisms are microscopically small sporozoans (single-celled organisms), *Plasmodium* spp., and the vectors (carriers) of the disease are certain anopheline mosquitos. More than 200 million people suffer from malaria every year, and of these more than two million die. Despite the past efforts of entomologists and public health officials, and their spectacular successes in eradicating malaria over considerable areas of South Africa, the disease remains a major problem in parts of the Transvaal and Natal. In fact, contrary to the euphoric expectations of earlier years when it was hoped to eradicate malaria worldwide, the disease is on the increase. This is due largely to the increasing resistance being shown by malaria parasites to most of the drugs currently available. (Mosquitos also transmit the filarial disease generally known as elephantiasis in which the lymphatic vessels of the body's extremities are blocked by nematodes.)

No-one likes lice, and perhaps with good reason. Like mosquitos, they transmit a range of diseases which result in suffering or death for many people. Regarding the prevalence of human lice in history, an amusing but illuminating account is given by Southwood (1977): 'When Thomas Beckett was undressed, following his murder at Canterbury in 1170, his innermost garments of haircloth seethed with lice like water simmering in a cauldron! Contemporary chroniclers regarded it as ... an additional sign of saintliness...' Typhus, a disease transmitted by lice, was the cause of a great number of human deaths in the 17th century, surpassed only by the plague as a cause of human mortality.

We have discussed only a few of the insects of medical importance here; a great many others exist which affect our health in some way or other, ranging from the sleeping sickness and leishmaniasis carried by tsetse flies and sand flies to the more common but painful stings inflicted by wasps and bees.

Insects in agriculture

An important way in which insects exert influence on humans is as agricultural pests. The world crop loss accredited to insects averages well in excess of 10 per cent during most years. This represents a vast amount of food. Man has turned to artificial means to attempt eradication, or at least control, of these pests and has brought the indirect importance of insects to the fore through the effect of insecticides on the general environment. One example is the classic case of the use of DDT.

The very effective insecticidal properties of DDT were first noticed during World War Two, when soldiers were protected from the bites of lice and fleas by the use of this chemical. After the war a tremendous surge in the production of DDT and similar compounds took place. The chemicals proved so effective in controlling agricultural pests and insects of medical importance that it raised food production to record levels, and for the first time in human history considerable hope existed that major scourges such as malaria could be wiped out by eradicating their insect carriers.

It almost worked; and there can by no denying that through the use of DDT the quality of life was raised for millions of people who would otherwise have suffered or died from disease and malnutrition. However, it was not long before people began noticing serious flaws in the properties and use of DDT and its related compounds. Two of the very properties which made it so highly effective, its high toxicity and its high stability (which caused it to remain in the environment for very long periods without its losing much of its poisonous effects), rendered it undesirable as a biocide, and people began realising that it was not the beneficial insecticide they had at first thought it to be. Although the chemical was applied to kill insect pests, it killed a distressing number of other animals as well. Another of the properties of DDT and similar compounds is that it is stored in the body, and this resulted in deaths in the longer term: for example, either the poison itself or insects which had died as a result of its ingestion were leached by rain and wind into rivers where small fish were living. The poison would accumulate in the bodies of these fish until it eventually killed them, or bigger fish would eat many of these smaller fish and so further concentrate the poison to higher levels. The bigger fish in turn would be eaten by predatory birds. All along this food chain, and in others, deaths resulted from the chlorinated hydrocarbon compounds, and eventually it was realised that they posed a threat to human life too: even if they didn't directly kill humans, there was evidence that the small amounts each person must inevitably consume could cause cancer. DDT and other related compounds are now banned in many countries.

Today we have narrower spectrum insecticides with relatively short residual lifespans, but even these have deleterious environmental effects. In many cases this is due to incorrect or overzealous dosages being applied or negligence, and it illustrates the point that a single irresponsible person brandishing a can of insecticide can inflict a lot of unnecessary damage and cause death.

Beneficial insects

Having painted a gloomy picture of insects, it would now be fair and correct to mention that they are also of incalculable value to man – indispensable, in fact. Insects contribute to a very large extent towards maintaining that balance between different organisms which is so essential in ensuring the continued existence of all species. To understand this point, one need only consider the importance of insects as pollinators: those beautiful flowers scattered in their millions around the world did not evolve so much to please our aesthetic sense as to ensure fertilization. The great majority of flowering plants depend heavily or exclusively on insect pollination, and without the assistance of insects these plants would soon die out – and as animals we in turn are dependent on these plants for food and oxygen.

Another example of how insects maintain the balance between different organisms is the considerable number of insect parasites, parasitoids and predators which feed on other insects. It is, in most cases, when we disrupt the effects of these beneficial organisms that we create pest conditions for ourselves. In nature there is at least one, but generally several, insect species feeding on every plant-feeding species. Odhiambo (1977) said that '...in tropical Africa... it has been thought that there might be well over 3 000 000 species [of insects]. Of that number, only about 0,3 per cent are major pests of man, his crops and his livestock...' A large number in the remaining 99,7 per cent aid us to the extent that without them we could not cope, by parasitizing and preying on potential pests so effectively that we do not even realize that they are 'pests'.

The study of insects benefits mankind in many ways, not the least of which is determining which species are pests and which are beneficial, and how we can aid and abet those insects which do help us.

Insects and the Kruger National Park

The Kruger National Park covers an area of nearly 20 000 square kilometres, is crossed by several large rivers and encompasses large open plains as well as hilly and mountainous areas. The average annual rainfall varies from about 744 mm in the Pretoriuskop area to 438 mm at Pafuri. This diversity of factors, together with other climatic and geological considerations, has a tremendous influence on the vegetation and, as is to be expected, the KNP supports a very rich and diverse insect fauna. The total number of known insect species in the KNP exceeds several thousand – and we have only just begun to scratch the surface. Prior to the mid-1970s entomological work in the KNP had been sporadic, with visiting scientists contributing much of what little knowledge we have of the ecological aspects of the smaller inhabitants of the reserve.

A magnificent reference collection has been built up over the years by staff members and has been added to from the collections of scientists who have worked in the area. This collection, based in Skukuza, owes its existence to Dr U de V Pienaar, who initiated and was the driving force behind the systematic survey of the insects of the KNP. The large and excellent butterfly collection was built up by Dr Pienaar, his family, and Johan Kloppers, with some contributions being made by other workers. Johan Kloppers, together with the late Dr G van Son, compiled a well illustrated book (1978) on the butterflies of the KNP.

The collecting and curating efforts of Harold and Toni Braack can be described only as magnificent, and have resulted in extensive reference material in various insect families, including the Scarabaeidae, Buprestidae, Cerambycidae (Coleoptera), Asilidae (Diptera), Noctuidae, Saturniidae, and Sphingidae (Lepidoptera). Miss HAD van Schalkwyk, formerly of the National Collection of Insects, during her retirement years spent countless hours classifying and identifying many of the insects in the Skukuza reference collection; it was due largely to her efforts that the collection is today organized in a manner which makes it a truly valuable reference source.

Visiting scientists who spent time working in the KNP and contributed significantly to our knowledge of local species include many well-known names such as GF Bornemissza (Scarabaeidae), HD Brown (Acrididae), WGH Coaton (Isoptera), M Coetzee (Culicidae), A Cornel (Culicidae), MC Ferreira (Scarabaeidae and Cerambycidae), E Holm (Buprestidae), IG Horak (insects of veterinary importance), YM Huang (Culicidae), R Hunt (Culicidae), M Mansell (Neuroptera), B de Meillon (Ceratopogonidae), R Meiswinkel (Ceratopogonidae), AJ Prins (Formicidae), C Scholtz (Trogidae), B and P Stuckenberg (Tabanidae), R Toms (Orthoptera), G van Eeden (Ceratopogonidae), L Vari (Lepidoptera), and F Zumpt (insects of medical and veterinary importance).

A reference collection can be of benefit and justify its existence only if it is actually being used, however, and the KNP's collection forms the basis of the very rich potential for extraordinary ecological work which could be done in this thriving wildlife reserve. It is in the relationship between insect and plant, or insect and animal, that the really significant dicoveries lie, and it is the elaboration on these relationships which lead to better conservation and management practices. We have an idea of what we have in the KNP, but we are only just starting to find out what they do and how they do it.

Insects of medical and veterinary importance in particular are at present receiving attention in the KNP because of the impact they have on wild animals and man through disease transmission and the debilitating effects their feeding and general harassment cause. This is not to say that the crucial role played by insects in plant fertilization, as well as other insect/plant interrelationships, do not remain of great concern, but an investigation of the role of insects in animal health accrues more immediate and visible benefits. There is a wealth of blood-sucking lice, bugs, sandflies, mosquitos, blackflies, biting midges, horseflies, blowflies, fleas and other potential or actual disease transmitters present in the KNP. The important question arises: how do they affect our wildlife?

Insects have already demonstrated their use as management tools in the KNP. Invasive exotic weeds such as *Pistia*, *Eichhornia* and *Salvinia* entering the KNP are now being selectively attacked by beetles adapted to feed on these plants, the beetles having been deliberately introduced to serve as biocontrol agents. In the same manner the moth *Cactoblastis cactorum* was introduced in an effort to control the spread of the exotic prickly pear *Opuntia ficus-indica*.

Communities of freshwater insects, such as the immatures of midges, damselflies and dragonflies, have been used as indicators of water quality; changes in the composition of these communities can indicate that river water is polluted. This aspect will become increasingly important due to dumping of mine effluent or leaching of agricultural fertilizers into rivers. Studies on the freshwater invertebrates – mainly insects – in the rivers of the KNP have provided baseline data which can be used for comparative purposes in future.

Key to Afrotropical of Orders of Hexapods (adults and immatures)

(Prepared by Prof DJ Brothers, University of Natal, Pietermaritzburg.)

This key has been designed in an attempt to enable the identification to order of any adult, larval or pupal hexapod collected in the Afrotropical region. However, it must be borne in mind that available information is so scanty that there are undoubtedly aberrant forms which have not been taken into account and which may 'key out' incorrectly; this is particularly true for young larvae and for groups where extrapolations have had to be made from information known only for other regions. Because of the variability within orders many couplets contain words such as 'usually', and the choice of one of the two options should then be made on the basis of the balance of probabilities.

I am grateful to the many colleagues who have commented on previous versions of this key. Their assistance is much appreciated.

1.	Wings well developed, at least one pair suitable for flight (forewing sometimes reduced and covering hind wing which is much folded and concealed at rest)	2
	Wings absent or reduced to pads or flap-like structures, not at all suitable for flight	39
2.(1)	Forewing horny, leathery or parchment-like, at least partially differing in texture from the membranous hind wing; forewing sometimes much shorter than hind wing (rarely vice versa)	3
	Forewing membranous, similar in texture to hind wing; hind wing sometimes reduced or absent	14
3.(2)	Mouthparts forming a tubular rostrum (sometimes very short) enclosing slender stylets for piercing and sucking	HEMIPTERA
	Mouthparts mandibulate, used for biting or chewing, or non-functional, without stylets	4
4.(3)	Forewing veinless although sometimes strongly sculptured, usually horny and heavily sclerotized, forming an elytron or pseudohaltere	5
	Forewing with definite veins, neither strongly sculptured nor horny, forming a tegmen	8
5.(4)	Forewing covering all or most of abdomen at rest	COLEOPTERA
	Forewing short, covering less than half of abdomen at rest	6

6.(5)		Antenna short, with seven or fewer segments, one or more segments each with a long lateral process; mouthparts reduced, non-functional; prothorax shorter than head .... STREPSIPTERA
		Antenna with 11 or more segments, usually simple but sometimes flabellate, mouthparts well-developed and functional; prothorax usually longer than head 7
7.(6)		Apex of abdomen with heavily sclerotized forceps; antenna long and slender with more than 11 segments; hind wing not entirely concealed by forewing, folded radially and crosswise .... DERMAPTERA
		Apex of abdomen without forceps; antenna usually 11-segmented, not very long and slender, often broadened apically; hind wing usually entirely concealed by forewing, if not then hind wing not folded crosswise but only radially .... COLEOPTERA
8.(4)		Hind wing similar in shape and size to forewing, without anal lobe; wings with humeral 'sutures' allowing them to be shed, leaving basal scales .... ISOPTERA
		Hind wing broader than forewing, with large anal lobe folded fanwise at rest; no humeral 'sutures' 9
9.(8)		Hind leg with femur very much longer than in other legs, generally saltatorial ... . ORTHOPTERA
		Hind leg with femur not markedly elongated, similar to middle leg, gressorial or cursorial 10
10.(9)		Foreleg highly modified, femur and tibia strikingly different from those of middle and hind legs 11
		All legs with femora and tibiae of basically similar form 12
11.(10)		Foreleg raptorial, elongate, femur and tibia subcylindrical and usually strongly spinose; head mobile on elongated (sometimes quadrate) pronotum .... MANTODEA
		Foreleg fossorial, stout, femur and tibia flattened with strong lamellate teeth; head not mobile, sunken into rounded pronotum .... ORTHOPTERA
12.(10)		Pronotum broad, flattened, shield-like, concealing most of head from above; all tibiae with many socketed spines; body of oval depressed form .... BLATTODEA
		Pronotum narrow, convex, not shield-like, not concealing most of head; tibiae without socketed spines; body usually subcylindrical (rarely strongly depressed) 13
13.(12)		Pronotum expanded posteromedially, longer than mesonotum, with a strong median carina; tarsi three-segmented .... ORTHOPTERA
		Pronotum quadrate, much shorter than mesonotum, without a strong median carina; tarsi five-segmented .... PHASMIDA
14.(2)		One (mesothoracic) pair of wings present and functional in flight (hind wing represented by a haltere or pseudohaltere, or absent) 15
		Two pairs of wings present, hind wing sometimes much smaller than forewing but of similar structure and functional in flight 17
15.(14)		Mouthparts well-developed, forming an evident proboscis (rarely reduced); hind wing represented by a club-like haltere; apex of abdomen without filamentous structures .... DIPTERA
		Mouthparts much reduced, non-functional; hind wing represented by a hook-like pseudohaltere or completely absent; apex of abdomen usually with filamentous structures 16
16.(15)		Wing with numerous longitudinal veins and crossveins; hind wing completely absent; antenna minute, setaceous .... EPHEMEROPTERA
		Wing with at most a single forked longitudinal vein, no crossveins; hind wing usually represented by a hook-like pseudohaltere; antenna long, filiform .... HEMIPTERA
17.(14)		Wings largely or entirely densely clothed with flattened, usually broad, scales on

	membrane and wings, without many long hairs on membrane; galea forming an elongated, coiled haustellum for sucking (rarely short or vestigial); mandible absent or vestigial (rarely well-developed and functional) . . . . LEPIDOPTERA Without dense, broad, flattened scales on wing membrane, rarely with some scales on veins (very rarely on membrane but then sparse and also with long hairs), wings sometimes with dense hairs or waxy dust; galea never forming a coiled haustellum; mandible usually present although sometimes concealed, sometimes vestigial or absent 18
18.(17)	Each tarsus with an apical eversible bladder, no claws; wings narrow and elongate, strap-like, with fringes of long hairs . . . . THYSANOPTERA Each tarsus without an eversible bladder, usually with claws and often with a pad-like arolium and/or pulvilli; wings not strap-like, usually without marginal fringes 19
19.(18)	Mouthparts forming a tubular rostrum (sometimes very short) enclosing slender stylets for piercing and sucking . . . . HEMIPTERA Mouthparts mandibulate, used for biting or chewing, or vestigial or absent, never with slender stylets 20
20.(19)	Foreleg raptorial, with coxa elongated and about as long as femur, unlike middle leg . . . . 21 Foreleg usually gressorial, with coxa not at all elongated and very much shorter than femur, basically similar to middle leg 22
21.(20)	Hind wing with broad anal lobe folded fanwise at rest; foretibia with strong hook-like apical spine extending beyond base of tarsus . . . . MANTODEA Hind wing without anal lobe; foretibia without apical hook-like spine . . . . NEUROPTERA
22.(20)	Hind leg with femur very much longer than in other legs, generally saltatorial . . . . ORTHOPTERA Hind leg with femur not markedly elongated, similar to middle leg, usually gressorial or cursorial, sometimes raptorial 23
23.(22)	Apex of abdomen with two or three elongated, multi-segmented filaments . . . . 24 Apex of abdomen without elongated, multi-segmented filaments, sometimes with long, unsegmented ovipositor and/or short cercus with up to three segments 25
24.(23)	Antenna minute, setaceous, much shorter than thorax; hind wing much shorter and usually narrower than forewing, without anal lobe . . . . EPHEMEROPTERA Antenna elongate, filiform, longer than head and thorax combined; hind wing about as long as forewing, with well-developed anal lobe and thus broader than forewing . . . . PLECOPTERA
25.(23)	Antenna minute, inconspicuous, setaceous, much shorter than thorax; wings not folded along abdomen at rest but held vertically or laterally . . . . ODONATA Antenna well-developed, obvious, usually at least as long as thorax; wings folded along abdomen at rest 26
26.(25)	Wings and body with a dense covering of fine hairs . . . . TRICHOPTERA Wings with membrane bare or at most with minute microtrichia; body usually not conspicuously hairy 27
27.(26)	Both pairs of wings with at least five crossveins in costal cell; general venation very complex, with many crossveins . . . . 28 Both pairs of wings with at most two crossveins in costal cell; general venation usually reduced, with few or no crossveins 30
28.(27)	Hind wing without defined anal area; ocelli absent (rarely present) . . . NEUROPTERA Hind wing with defined anal area folded under remigium at rest; ocelli present (rarely absent) 29
29.(28)	Tarsi five-segmented . . . . MEGALOPTERA Tarsi three-segmented . . . . PLECOPTERA

30.(27)	Apparent thorax and apparent abdomen strongly differentiated by a marked constriction, allowing considerable movement	HYMENOPTERA
	Abdomen broadly attached to thorax, not strongly constricted basally, not independently movable	31
31.(30)	Head produced ventrally as a beak-like structure with well-developed, elongate mandible and labrum; tarsi raptorial, each with a single stout claw	MECOPTERA
	Head without ventral beak-like projection; tarsi usually simple, each with paired claws	32
32.(31)	Tarsi four-segmented; wings subequal, with well-developed humeral 'sutures', allowing them to be shed leaving basal scales	ISOPTERA
	Tarsi five-, three- or two-segmented; forewing usually longer than hind wing; wings without definite humeral 'sutures' (very rarely with indefinite basal fractures but then forewing always much larger than hind wing)	33
33.(32)	Tarsi five-segmented	34
	Tarsi three- or two-segmented	36
34.(33)	Forewing with pterostigma well developed, heavily sclerotized and distinctly bounded	HYMENOPTERA
	Forewing usually lacking pterostigma; if present then neither more heavily sclerotized than wing membrane nor distinctly bounded	35
35.(34)	Forewing with vein Rs at least three-branched; tibiae with conspicuous mobile spurs; body and wings without waxy dust; thoracic pleura evenly sclerotized	TRICHOPTERA
	Forewing with vein Rs two-branched; tibiae without conspicuous spurs; body and wings covered with waxy dust; thoracic pleura apparently largely membranous but with conspicuous sclerotised bars	NEUROPTERA
36.(33)	Basal segment of foretarsus conspicuously swollen, bearing silk ejectors on ventral surface; genitalia conspicuous and highly asymmetrical	EMBIIDINA
	Basal segment of foretarsus not at all swollen, without silk-producing organs; genitalia inconspicuous, usually symmetrical	37
37.(36)	Hind wing with broad anal lobe folded under remigium at rest, thus broader than forewing; wings with numerous parallel longitudinal veins apically	PLECOPTERA
	Hind wing without anal lobe, narrower than forewing; wings with very few longitudinal veins apically	38
38.(37)	Apex of abdomen with short cerci; pronotum well developed, at least as long as mesonotum; antenna moniliform, nine-segmented	ZORAPTERA
	Apex of abdomen without cerci; pronotum much reduced, very much shorter than mesonotum; antenna filiform, with 11 or more segments (very rarely fewer segments)	PSOCOPTERA
39.(1)	Three pairs of thoracic legs present, usually segmented, although sometimes vestigial, inconspicuous or closely appressed to body	40
	Thoracic legs entirely absent (rarely paired or unpaired ventral processes present on prothorax only)	100
40.(39)	Labium forming large, elongate, extensile mask for prey capture; labial palp unsegmented, expanded and with at least one strong, socketed spine, or hook apically	ODONATA
	Labium usually small and inconspicuous, sometimes elongate and enclosing stylets; labial palp unspecialized or absent	41
41.(40)	Lamellate or filamentous gills present along abdomen and/or on thorax; aquatic (very rarely terrestrial)	42
	Gills entirely absent (abdomen sometimes with inconspicuous ventral styles); usually terrestrial, sometimes parasitic or aquatic but then abdomen never with styles	49
42.(41)	Apex of abdomen with two or three multi-segmented filaments	43
	Apex of abdomen without multi-segmented filaments; sometimes with anal gills	

	and/or prolegs or processes 44	
43.(42)	Gills along sides of abdomen, usually lamellate, none on thorax; abdominal apex with three filaments, rarely two	EPHEMEROPTERA
	Gills forming tufts of filaments on thorax and between cerci, none along sides of abdomen; abdominal apex with two filaments	PLECOPTERA
44.(42)	Mandible and maxilla closely associated, forming a slender, straight sucking tube much longer than head	NEUROPTERA
	Mandible not closely associated with maxilla, strongly curved, usually shorter than head 46	
46.(45)	Mandible used for biting; antenna lateral to anterior arm of ecdysial line; apex of abdomen extended as an unsegmented median filament	MEGALOPTERA
	Mandible used for piercing and sucking; antenna mesal to anterior arm of ecdysial line; abdominal apex not filamentous	COLEOPTERA
47.(44)	Labrum absent; antenna mesal to anterior arm of ecdysial line	COLEOPTERA
	Labrum present although sometimes retracted under clypeus; antenna lateral to anterior arm of ecdysial line 48	
48.(47)	Antenna unsegmented, usually minute; no abdominal spiracles; abdominal gills usually ventrally oriented; usually found within a portable case or fixed retreat constructed of silk and often covered with sand grains or pieces of vegetation	TRICHOPTERA
	Antenna four-segmented, obvious; eight pairs of abdominal spiracles; abdominal gills laterally oriented; never found within a silken case or retreat . . .	MEGALOPTERA
49.(41)	Apex of abdomen with heavily sclerotized forceps 50	
	Apex of abdomen without forceps, sometimes with simple unsegmented to multi-segmented appendages 51	
50.(49)	Eyes absent; tarsi unsegmented; body almost entirely pale and unsclerotized	DIPLURA
	Eyes present; tarsi three-segmented; body dark, sclerotized	DERMAPTERA
51.(49)	First abdominal segment with median ventral tube; abdomen with six segments of which some may be fused; fourth abdominal sternum usually with forked springing organ, this sometimes reduced or absent	COLLEMBOLA
	First abdominal segment without median ventral tube; abdomen with eight or more segments distinguishable, segmentation rarely indistinct or apparently reduced; abdominal springing organ never present 52	
52.(51)	Apex of abdomen with three multi-segmented filamentous appendages 53	
	Apex of abdomen with two or no multi-segmented filaments (paired short appendages sometimes present) 54	
53.(52)	Compound eye large, holoptic; cercus shorter than median filament; body subcylindrical	ARCHAEOGNATHA
	Compound eye small, dichoptic, or absent; cercus subequal to median filament; body dorso-ventrally depressed	THYSANURA
54.(52)	Anterior abdominal sterna with paired, unsegmented or two-segmented styles; compound eyes and ocelli absent; mouthparts entognathous 55	
	Anterior abdominal sterna without styles (paired styles rarely present toward apex of abdomen); compound eyes, stemmata and/or ocelli usually present; mouthparts ectognathous (sometimes concealed when suctorial) 56	
55.(54)	Antenna absent; apex of abdomen simple, without appendages	PROTURA
	Antenna well-developed, elongate, moniliform; apex of abdomen with paired, multi-segmented cerci	DIPLURA
56.(54)	Abdomen with paired prolegs or similar ventral processes which often bear sclerotized hooks 57	
	Abdomen without paired prolegs or processes, sometimes with a single apical proleg or pseudopod, usually without sclerotized hooks 63	
57.(56)	Abdomen with a single pair of prolegs or processes, at its apex 58	

	Abdomen with two or more pairs of prolegs or similar processes	59
58.(57)	Antenna multi-segmented, well-developed; mandible and maxilla coadapted as an elongate sucking tube	NEUROPTERA
	Antenna unsegmented, usually minute; mandible and maxilla separate, used for biting and chewing	TRICHOPTERA
59.(57)	Abdominal prolegs each with one or more (often many) delicate, sclerotized hooks (crochets)	LEPIDOPTERA
	Abdominal prolegs without crochets	60
60.(59)	Fewer than five pairs of abdominal prolegs	COLEOPTERA
	More than five pairs of abdominal prolegs	61
61.(60)	A single stemma on each side of head, sometimes none.	HYMENOPTERA
	Few to many stemmata on each side of head	62
62.(61)	Many stemmata, strongly grouped, apparently forming a compound eye; a median 'ocellus' present	MECOPTERA
	Few stemmata, distinctly separated by cuticle, unlike a compound eye; no median 'ocellus'	LEPIDOPTERA
63.(56)	Larva, nymph or imago (adult), active; legs capable of movement	64
	Pupa, usually inactive; legs usually incapable of movement	111
64.(63)	Body strongly laterally compressed; ectoparasitic on mammals or birds	SIPHONAPTERA
	Body subcylindrical or dorso-ventrally depressed; usually free-living (if ectoparasitic on mammals or birds, then depressed)	65
65.(64)	Hind leg with femur and tibia very much longer than in other legs, generally saltatorial	ORTHOPTERA
	Hind leg with femur and tibia not markedly elongated, similar to middle leg although sometimes with thicker femur, generally not saltatorial	66
66.(65)	Mandible and maxilla closely coadapted to form a piercing and sucking tube at each side of head	NEUROPTERA
	Mandible and maxilla separate or vestigial or absent (if mouthparts are suctorial then a median proboscis is developed or mandible alone is modified and grooved or tubular)	67
67.(66)	Apex of abdomen without cerci or similar paired appendages visible externally (small cerci sometimes concealed between apical sclerites)	68
	Apex of abdomen with cerci, styles or similar unsegmented or segmented paired appendages visible externally	84
68.(67)	Mouthparts with well-developed mandible of a biting or chewing type, rarely working outwards	69
	Mouthparts suctorial or atrophied, if mandible functional then styliform	76
69.(68)	Body strongly dorso-ventrally depressed, ectoparasitic on mammals or birds	MALLOPHAGA
	Body subcylindrical or rounded, not markedly depressed; not ectoparasitic on mammals or birds	70
70.(69)	Apparent thorax and apparent abdomen strongly differentiated by a marked constriction, allowing considerable movement	HYMENOPTERA
	Abdomen broadly attached to thorax, not strongly constricted basally, not independently movable	71
71.(70)	Body elongate, eruciform; legs short; thorax and abdomen not well differentiated, often without distinct sclerites	72
	Body compact, usually not elongate, not eruciform; legs relatively long; thorax and abdomen usually distinct and with separate sclerites	74
72.(71)	Labium without a spinneret; usually more than one stemma on each side of head, sometimes one or none	COLEOPTERA
	Labial spinneret present; usually one or no stemmata on each side	73
73.(72)	Found in leaf mine; prothorax with sclerotized dorsal shield	LEPIDOPTERA

	Found in burrow in wood or on plant surface (rarely in leaf mine); prothorax without sclerotized dorsal shield .... HYMENOPTERA
74.(71)	Clypeus divided into transverse anteclypeus and swollen postclypeus; pronotum much reduced, very much shorter than mesonotum; antenna with 12 or more segments (rarely fewer) .... PSOCOPTERA
	Clypeus simple, undivided, not swollen; pronotum not much reduced, about as long as mesonotum or longer; antenna with 11 segments or fewer 75
75.(74)	Legs with trochanters .... COLEOPTERA
	Legs without trochanters .... STREPSIPTERA
76.(68)	Legs without trochanters; mouthparts atrophied .... STREPSIPTERA
	Legs with trochanters; mouthparts usually well developed although sometimes not obvious externally, sometimes atrophied 77
77.(76)	Body dorso-ventrally depressed; ectoparasitic on mammals, birds or bees ... 78
	Body subcylindrical or rounded; usually free-living (always free-living if body somewhat depressed) 80
78.(77)	Each tarsus with a single, large claw closing against an inner apical process of the tibia; body poorly sclerotized .... ANOPLURA
	Each tarsus with normal paired claws or none and tibia without an inner apical process; body usually well sclerotized 79
79.(78)	Antenna conspicuous, elongate; mouthparts forming a segmented rostrum enclosing slender stylets; pronotum well developed, subequal to mesonotum .... HEMIPTERA
	Antenna minute, subglobose; mouthparts usually forming a short, unsegmented proboscis; pronotum much reduced, much shorter than mesonotum .... DIPTERA
80.(77)	Body densely clothed with scales and hairs; mouthparts forming an elongate, coiled haustellum, sometimes much reduced .... LEPIDOPTERA
	Body without dense scales, sometimes with hairs; mouthparts never forming a coiled haustellum 81
81.(80)	Tarsi unsegmented to three-segmented; mouthparts forming a sucking tube or cone which may be short but which encloses stylets 82
	Tarsi five-segmented; mouthparts not forming a tube or cone enclosing stylets 83
82.(81)	Each tarsus with an apical, eversible bladder, no claws; mouthparts asymmetrical, a cone enclosing three short stylets .... THYSANOPTERA
	Each tarsus without an eversible bladder, with paired claws; mouthparts symmetrical, an elongate tubular rostrum (sometimes short) enclosing four long, slender stylets .... HEMIPTERA
83.(81)	Vestiges of two pairs of wings present; antenna much longer than thorax .... TRICHOPTERA
	Vestiges of only one pair of wings or no trace of wings (halteres sometimes present); antenna usually shorter than thorax .... DIPTERA
84.(67)	Mandible absent or vestigial (rarely functional, but then styliform and associated with suctorial mouthparts) 85
	Mandible well developed, adapted for biting and/or chewing 87
85.(84)	Antenna absent; legs without trochanters; tarsi unsegmented ...... STREPSIPTERA
	Antenna present; legs with trochanters; tarsi four- or five-segmented 86
86.(85)	Head capsule heavily sclerotized, with pronounced anterior snout-like projection; eyes absent; tarsi four-segmented .... ISOPTERA
	Head capsule not unusually heavily sclerotized, without any anterior snout-like projection; eyes present; tarsi five-segmented .... DIPTERA
87.(84)	Foreleg highly modified, femur and tibia strikingly different from those of middle and hind legs 88
	All legs with femora and tibiae of basically similar form 89
88.(87)	Foreleg raptorial, elongate, femur and tibia subcylindrical and usually strongly

	spinose; head mobile on elongated (sometimes quadrate) pronotum . . MANTODEA
	Foreleg fossorial, stout, femur and tibia flattened with strong lamellate teeth; head not mobile, sunken into rounded pronotum . . . . ORTHOPTERA
89.(87)	Apparent thorax and apparent abdomen strongly differentiated by a marked constriction, allowing considerable movement . . . . HYMENOPTERA
	Abdomen broadly attached to thorax, not strongly constricted basally, not independently movable 90
90.(89)	Basal segment of foretarsus conspicuously swollen, bearing silk ejectors on ventral surface . . . . EMBIIDINA
	Basal segment of foretarsus not at all swollen, without silk-producing organs . . 91
91.(90)	Tarsi unsegmented, each usually with a single claw; thorax and abdomen usually poorly differentiated dorsally . . . . COLEOPTERA
	Tarsi two- to five-segmented, each usually with paired claws; thorax and abdomen usually distinct dorsally 92
92.(91)	Tarsi two-segmented; antenna nine-segmented . . . . ZORAPTERA
	Tarsi three- to five-segmented; antenna usually with more than nine segments 93
93.(92)	Cercus very long, about as long as abdomen or longer; tarsi three-segmented 94
	Cercus short, at most half the length of abdomen, usually much shorter; tarsi four- or five-segmented, sometimes three-segmented 95
94.(93)	Ocelli present; cercus multi-segmented; aquatic . . . . PLECOPTERA
	Ocelli absent; cercus usually unsegmented, sometimes multi-segmented; terrestrial, sometimes closely associated with rodents . . . . DERMAPTERA
95.(93)	Pronotum broad, flattened, shield-like, concealing most of head from above; all tibiae with many socketed spines; body of oval depressed form . . . . . . BLATTODEA
	Pronotum not concealing any part of head from above (rarely slightly concealing posterior part of head but then not at all flattened); tibiae without socketed spines; body usually subcylindrical or rounded (rarely strongly depressed) . . . 96
96.(95)	Tarsi raptorial, each with a single, stout claw; head ventrally produced as a beak-like structure with well-developed elongate mandible and labrum . . . . . MECOPTERA
	Tarsi simple, each with paired claws; head without ventral beak-like projection (if with snout-like projection then this is anterior and distant from small mandible) 97
97.(96)	Pronotum less than half the length of mesonotum; body greatly elongated and stick-like or grass-like, rarely very flattened and leaf-like . . . . PHASMIDA
	Pronotum about as long as or longer than mesonotum; body more or less compact, not stick-like, grass-like or leaf-like 98
98.(97)	Tarsi four-segmented; compound eyes usually absent, if present then wings almost always represented by scale-like bases . . . . ISOPTERA
	Tarsi three-segmented; compound eyes always present; wings never represented by scale-like bases, sometimes present as wing pads 99
99.(98)	Ocelli absent; pronotum flattened; cercus prominent . . . . DERMAPTERA
	Ocelli present; pronotum highly convex with median ridge; cercus inconspicuous . . . . ORTHOPTERA
100.(39)	Strongly flattened (sometimes only ventrally) and found closely appressed to host plant, covered by waxy secretion forming scale, powdery particles or filaments; mouthparts long, slender stylets, functional (rarely non-functional) . . . . HEMIPTERA
	Subcylindrical or rounded (rarely flattened ventrally), not usually found closely appressed to vegetation, never covered by waxy secretions; mouthparts never long, slender stylets, sometimes non-functional 101
101.(100)	Thin-walled anal papillae and/or tracheal gills present on abdomen; aquatic . . . . DIPTERA
	Anal papillae and tracheal gills absent; usually terrestrial although sometimes

	parasitic 102
102.(101)	Mandible well developed, usually of a biting or chewing type; mouthparts functional; head capsule well defined although sometimes not sclerotized, often more or less hemispherical 103
	Mandible reduced or absent, sometimes replaced by vertical hoods; mouthparts sometimes non-functional; no well-defined head capsule, head usually soft and often more or less conical, sometimes fused with thorax 108
103.(102)	Prothorax broader than other body segments, distinct . . . . COLEOPTERA
	Prothorax narrower than at least one other body segment or of equal width (all thoracic segments sometimes fused and broader than abdomen) 104
104.(103)	Head prognathous; body usually markedly elongate 105
	Head hypognathous (rarely tending towards prognathous); body fairly compact, rarely markedly 106
105.(104)	Body with 13 segments, each with a ring of prominent setae; apex of abdomen with a pair of unsegmented anal struts; antenna prominent, unsegmented . . . . SIPHONAPTERA
	Body with 12 or fewer segments (rarely apparently more than 12 because of subdivision), usually without a ring of setae on each segment; apex of abdomen usually without simple unsegmented appendages; antenna usually inconspicuous . . . . DIPTERA
106.(104)	Body elongate, more or less straight and of constant thickness; antenna lateral to anterior arm of ecdysial line . . . . LEPIDOPTERA
	Body fairly compact, thickest in the middle and/or curved dorsoventrally (rarely elongate); antenna mesal (rarely lateral) to anterior arm of ecdysial line or anterior arm absent or not reaching level of antenna 107
107.(106)	Abdominal terga each with three or four transverse folds; head capsule usually sclerotized, dark; labial spinneret absent . . . . COLEOPTERA
	Abdominal terga with at most two transverse folds, usually none, although sometimes with tubercles; head unsclerotized, pale; labial spinneret present . . . . HYMENOPTERA
108.(102)	Mouthparts functional, often comprising a pair of internal vertical mouth hooks . . . . DIPTERA
	Mouthparts vestigial or absent, non-functional 109
109.(108)	Thorax and abdomen well-differentiated and distinct; endoparasitic in other insects, sometimes with fused head and thorax protruding between abdominal sclerites of living host . . . . STREPSIPTERA
	Thorax and abdomen not well differentiated; free-living (rarely within remains of dead host) 110
110.(109)	Body poorly sclerotized; at least seven pairs of spiracles along abdomen; found within a bag constructed of silk and pieces of vegetation . . . . LEPIDOPTERA
	Body with well-sclerotized outer layer forming a puparium enclosing the exarate pupa; no spiracles along abdomen (at most one pair at its apex); not found within a silken bag . . . . DIPTERA
111.(63)	Apparent thorax and apparent abdomen differentiated by a marked constriction . . . . HYMENOPTERA
	Abdomen broadly attached to thorax, not markedly constricted basally 112
112.(111)	Mandible well-developed, distinct, of a basic biting or chewing type although sometimes not functional 113
	Mandible usually absent, rarely rudimentary or styliform, never of a basic biting or chewing type 120
113.(112)	Mandible immovable (adecticous), usually not heavily sclerotized 114
	Mandible movable (decticous), usually heavily sclerotized 115
114.(113)	Pronotum much smaller than mesonotum and poorly differentiated from it . . . . HYMENOPTERA

19

	Pronotum at least as large as mesonotum, often larger, well differentiated from it .... COLEOPTERA
115.(113)	Head ventrally produced as a beak-like structure with elongate mandible and labrum 116
	Head not ventrally produced, not beak-like or snout-like, mandible usually not markedly elongated and labrum transverse 117
116.(115)	Hind wing pad about as long as forewing pad or shorter, straight; legs, especially tarsi, markedly elongated .... MECOPTERA
	Hind wing pad very much longer than forewing pad, spirally coiled apically; legs not unusually elongated .... NEUROPTERA
117.(115)	Pronotum usually reduced and inconspicuous, shorter than mesonotum; mandible transverse; body usually straight 118
	Pronotum well developed and obvious, usually at least as long as mesonotum; mandible usually more or less longitudinal or oblique; head and apex of abdomen usually deflexed ventrally 119
118.(117)	Pronotum much shorter than mesonotum; middle leg often with fringe of hairs for swimming; gills sometimes present; aquatic (very rarely terrestrial) .... TRICHOPTERA
	Pronotum slightly shorter than mesonotum; middle leg without fringe of hairs; gills never present; always terrestrial .... LEPIDOPTERA
119.(117)	Found enclosed within a cocoon constructed of silk and often with soil particles incorporated .... NEUROPTERA
	Found within a simple cell in the soil, not in a cocoon .... MEGALOPTERA
120.(112)	Appendages free, not cemented to body (exarate) 121
	Appendages cemented to body (obtect) 123
121.(120)	Antenna minute, scarcely distinguishable; body laterally compressed; tarsi elongated .... SIPHONAPTERA
	Antenna obvious, usually elongate; body subcylindrical; tarsi short 122
122.(121)	Body strongly elongate, abdomen much longer than thorax; two pairs of conspicuous wing pads about equal in size, or none .... THYSANOPTERA
	Body more or less compact, abdomen no longer than thorax; one pair of well developed wing pads (hind wing pad absent) .... HEMIPTERA
123.(120)	Two pairs of wing pads (rarely none); mouthparts usually very long and obvious; abdominal spiracles functional; no anterior respiratory horns .... LEPIDOPTERA
	One pair of wing pads (rarely none); mouthparts usually short and inconspicuous; abdominal spiracles usually non-functional; anterior respiratory horns often present .... DIPTERA

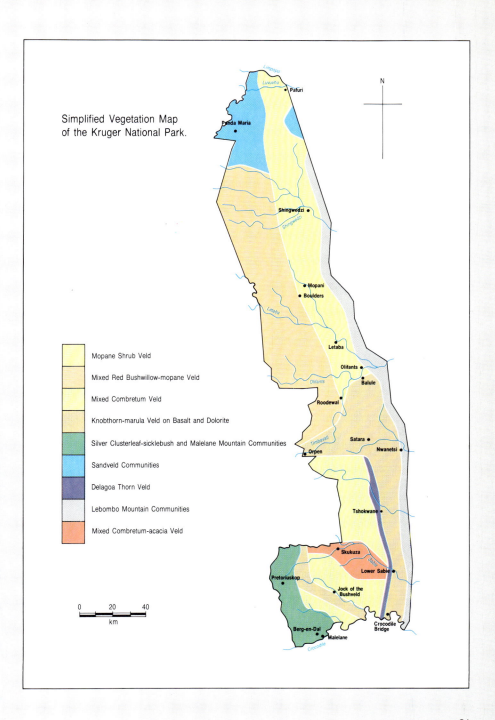

THYSANURA / EPHEMEROPTERA / ODONATA

Silverfish (page 65)

Mayfly (page 66)

(Anisoptera) adult dragonfly (page 66)

(Zygoptera) damselflies mating (page 66)

(Anisoptera) aquatic nymph of dragonfly (page 66)

PLATE 1

PHASMATODEA / MANTODEA / BLATTODEA

Stick insect (page 68)

(Empusidae) praying mantid (page 68)

(Hymenopodidae) praying mantid *Pseudocreobotra wahlbergi* (page 68)

Cockroach (page 69)

PLATE 2

ISOPTERA / ORTHOPTERA

(Termitidae) mound of *Macrotermes natalensis* (page 72)

(Termitidae) male and female reproductive termites (page 70), having just lost their wings after swarming, scurry away in tandem

(Termitidae, Nasutitermitinae) snouted termites (page 73)

(Acrididae) grasshopper (page 74)

(Pyrgomorphidae) *Zonocerus elegans* (page 75)

PLATE 3

ORTHOPTERA

(Tettigoniidae) katydid (page 76)

(Tettigoniidae, Hetrodinae) armoured ground cricket (page 76)

(Gryllidae) cricket *Brachytrupes membranaceus* (page 76)

(Gryllotalpidae) mole cricket (page 77)

PLATE 4

DERMAPTERA / EMBIIDINA / PLECOPTERA / PSOCOPTERA

Booklouse (page 78)

Web-spinner (page 78)

Stonefly (page 78)

Earwig (page 77)

PLATE 5

PHTHIRAPTERA

(Mallophaga) biting louse (page 80)

(Anoplura) sucking lice *Haematopinus phacocoeri* (page 81)

(Rhyncophthirina) elephant lice *Haematomyzus elephantis* (page 80)

PLATE 6

HEMIPTERA

(Cicadidae) cicadas or Christmas beetles (page 82)

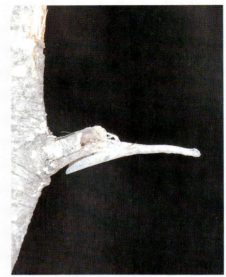

(Fulgoridae) lantern fly (page 83)

(Cercopidae) immature spittle bugs (page 83) concealed within their frothy secretion

(Cicadellidae) leafhopper (page 83)

(Cercopidae) adult spittle bugs (page 83)

PLATE 7

HEMIPTERA

(Membracidae) treehopper (page 84)

(Psyllidae) immatures of *Arytaina mopani* (page 84) with their protective waxy scales removed

(Psyllidae) adult *Arytaina mopani* (page 84)

(Psyllidae) leaf-curl of mopane trees, especially prevalent during the early rainy season, caused by *Arytaina mopani* (page 84)

(Aphididae) aphids or plantlice (page 85)

PLATE 8

HEMIPTERA

(Coccidae) soft scale insects (page 85) excrete sweet substances which are avidly sought after by ants

(Cydnidae) burrowing bug (page 86)

(Coreidae) squash bug (page 86)

(Coreidae) spiny squash bug *Pephricus livingstonei* (page 86)

(Lygaeidae) seed bugs (page 87)

PLATE 9

HEMIPTERA

(Pentatomidae) shield bug (page 86) feeding on a caterpillar

(Pyrrhocoridae) cotton stainers (page 87)

(Reduviidae) assassin bugs (page 87)

(Cimicidae) bat bugs (page 87)

(Miridae) plant bug (page 88)

PLATE 10

HEMIPTERA

(Gerridae) water striders mating (page 88)

(Notonectidae) backswimmer (page 88)

(Corixidae) water-boatman (page 89)

(Nepidae) water-scorpion (page 89)

PLATE 11

HEMIPTERA / THYSANOPTERA / MEGALOPTERA

(Belostomatidae) giant water bug (page 89)

Thrips (page 90)

Alderfly (page 91)

PLATE 12

NEUROPTERA

(Chrysopidae) lacewing (page 91)

(Mantispidae) mantispid (page 92)

(Myrmeleontidae) antlion pits and tracks, made by immature stages (page 92)

(Myrmeleontidae) antlion adults (page 92)

PLATE 13

NEUROPTERA / SIPHONAPTERA

(Ascalaphidae) owl fly (page 93)

Sticktight fleas *Echidnophaga larina* (page 94) with mouthparts embedded in a warthog

Bat flea *Lagaropsylla idae* (page 95)

PLATE 14

COLEOPTERA

(Carabidae) ground beetle *Anthia* sp. (page 96)

(Cicindelidae) tiger beetle (page 96) with a captured fly

(Paussidae) paussid (page 97)

(Gyrinidae) whirligig beetle (page 97)

(Dytiscidae) diving beetle (page 97)

PLATE 15

COLEOPTERA

(Histeridae) histerid beetles (page 98)

(Staphylinidae) rove beetle (page 98)

(Lycidae) net-winged beetle (page 99)

(Lampyridae) adult, wingless, female glow-worm (page 99)

(Lampyridae) adult, winged, male firefly (page 99)

PLATE 16

COLEOPTERA

(Cleridae) checkered beetles *Necrobia rufipes* (page 99)

(Lymexylidae) lymexylid beetle (page 100)

(Elateridae) click beetle (page 100)

COLEOPTERA

(Buprestidae) metallic woodborer (page 100)

(Coccinellidae) ladybird beetle (page 101)

(Dermestidae) skin beetles *Dermestes maculatus* (page 101) on a carcass

PLATE 18

COLEOPTERA

(Bostrychidae) shot-hole borer (page 101)

(Meloidae) blister beetle (page 102)

(Tenebrionidae) darkling beetle (page 102)

(Scarabaeidae, Scarabaeinae) dung beetle (page 103)

PLATE 19

COLEOPTERA

(Scarabaeidae, Dynastinae) rhinoceros beetle (page 104)

(Scarabaeidae, Cetoniinae) flower chafer (page 104)

(Trogidae) trogid beetle (page 105)

(Scarabaeidae, Rutelinae) leaf chafer (page 105)

PLATE 20

COLEOPTERA

(Cerambycidae) adult long-horn beetle (page 105)

(Cerambycidae) mabungu grubs (larval *Macrotoma* sp.) (page 105)

COLEOPTERA

(Chrysomelidae, Cassidinae) tortoise beetle (page 106)

(Chrysomelidae, Alticinae) leaf beetle (page 106)

(Bruchidae) seed weevil (page 106)

(Scolytidae) breeding chambers of bark beetles (page 106)

(Curculionidae) snout weevil (page 107)

PLATE 22

COLEOPTERA / STREPSIPTERA / MECOPTERA

(Brentidae) straight-snouted weevil (page 108)

Stylopid (page 108)

Hanging fly (page 108)

DIPTERA

(Tipulidae) crane fly (page 109)

(Culicidae) *Mansonia* sp. (page 110) feeding on human host

(Psychodidae) moth-fly (page 110)

(Ceratopogonidae) adult biting midge *Culicoides imicola* (page 112), one of the most important transmitters of arthropod-borne viruses in South Africa

(Culicidae) mosquito larva *Anopheles* sp. (page 110)

PLATE 24

DIPTERA

(Chironomidae) midge (page 112)

(Cecidomyiidae) numerous adult gall midges (page 113) resting on silk strands within a tree-hole

(Cecidomyiidae) galls on a twig of silverleaf terminalia caused by the gall midge *Tetrasphondylia terminaliae* (page 113)

(Tabanidae) horse fly (page 113)

DIPTERA

(Asilidae) robber fly (page 114)

(Bombyliidae) bee fly (page 114)

(Syrphidae) hover fly (page 115)

DIPTERA

(Lauxaniidae) (page 115)

(Muscidae) *Rhinomusca dutoiti* (page 116) feeding at rhino lesion

(Gasterophilidae) larval *Gasterophilus* (page 117) attached to stomach of zebra

(Oestridae) larval bot flies (*Oestrus* sp. and *Gedoelstia* sp.) (page 117) in the skull of a reedbuck

DIPTERA

(Calliphoridae) blowfly *Chrysomya albiceps* (page 118) on an elephant carcass

(Calliphoridae) densely packed blowfly larvae (page 118) at a carcass

(Sarcophagidae) flesh fly (page 120)

PLATE 28

DIPTERA

(Tachinidae) (page 120)

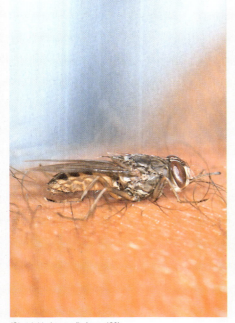

(Glossinidae) tsetse fly (page 120)

(Hippoboscidae) wingless, adult louse fly, *Lipoptena paradoxa* (page 121) resting on bushbuck

(Hippoboscidae) louse fly (page 121)

PLATE 29

DIPTERA / TRICHOPTERA

(Diopsidae) stalk-eyed fly (page 122)

(Platystomatidae) redheaded fly *Bromophila caffra* (page 122)

(Conopidae) thick-headed fly (page 122)

(Piophilidae) skipper fly *Piophila casei* (page 123)

Caddisfly (page 123)

PLATE 30

LEPIDOPTERA

(Psychidae) protective larval case characteristic of bagworms (page 124)

(Tineidae) larval tubes of *Ceratophaga vastella* (page 125)

(Zygaenidae) burnet *Zutulba zelleri* (page 125)

(Pyralidae) *Cactoblastis cactorum* (page 126)

(Pterophoridae) plume moth (page 126)

LEPIDOPTERA

(Geometridae) looper or earth measurer (page 127)

(Saturniidae) emperor moth *Argema mimosae* (page 128)

(Sphingidae) hawk moth or sphinx moth (page 128)

PLATE 32

LEPIDOPTERA

(Thaumetopoeidae) processionary moth *Anaphe reticulata* (page 129)

(Noctuidae) army worm *Spodoptera exempta* (page 130) in the dark phase

(Thaumetopoeidae) column of processionary worms *Anaphe reticulata* (page 129)

(Noctuidae) owl moth *Cyligramma latona* (page 129)

PLATE 33

LEPIDOPTERA

(Ctenuchidae) handmaiden (page 130)

(Arctiidae) tiger moth *Utetheisa pulchella* (page 130)

PLATE 34

LEPIDOPTERA

(Papilionidae) caterpillar of *Papilio demodocus* (page 131) with extruded osmeteria

(Papilionidae) swallowtail *Papilio demodocus* (page 131)

PLATE 35

LEPIDOPTERA

(Hesperiidae) skipper (page 131)

(Lycaenidae) blue (page 132)

(Pieridae) white *Belenois aurota* (page 132)

PLATE 36

LEPIDOPTERA

(Charaxidae) *Charaxes brutus* (page 133)

(Nymphalidae) brush-footed butterflies (page 133)

LEPIDOPTERA

(Acraeidae) red (page 134)

(Satyridae) brown (page 135)

(Danaidae) African monarch butterfly *Danaus chrysippus* (page 134)

PLATE 38

HYMENOPTERA

(Tenthredinidae) sawfly (page 136)

(Chalcidoidea) chalcid wasp *Pteromalus* sp. (page 137) parasitising a cocoon of a braconid wasp

(Ichneumonidae) (page 136)

HYMENOPTERA

(Chrysididae) cuckoo wasp (page 138)

(Mutillidae) velvet ants (page 138)

(Pompilidae) spider-hunting wasp (page 139)

(Eumenidae) potter wasp or mason wasp (page 139)

PLATE 40

HYMENOPTERA

(Vespidae) social wasp (page 139)

(Sphecidae) (page 140)

(Megachilidae) leafcutter bee (page 140)

PLATE 41

HYMENOPTERA

(Xylocopidae) carpenter bee (page 141)

(Apidae) honeybee (page 141)

PLATE 42

HYMENOPTERA

(Formicidae) winged, reproductive male of the driver ant *Dorylus* sp. (page 144)

(Formicidae) column of foraging Matabele ants *Megaponera foetens* (page 143)

PLATE 43

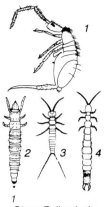

1
*Class Collembola:
Arthropleona*
2
Class Protura
3
*Class Diplura:
Campodeidae*
4
Diplura: Japygidae

PLATE 1

CLASSES PROTURA, DIPLURA and COLLEMBOLA

Although falling within the domain of entomologists, the small, soft-bodied animals of these three classes have their mouthparts situated within the buccal cavity and are therefore by strict definition not part of the class Insecta. As they are six-legged arthropods they are often mistaken for true insects, however. All three classes are widely distributed on most continents. In the KNP no effort has been made to collect species from these groups, although representatives of each were retrieved from leaf-litter samples taken by the author along the Sabie River near Skukuza.

The Protura generally measure less than 2,5 mm long, are elongate, and have pincer-like or fairly long, multi-segmented cerci posteriorly. They are common in most environments where moist soil or fairly thick leaf-litter occurs, and may be found under stones or logs. The Diplura are slightly larger (and some, exceptional, species reach nearly 50 mm in length) and are also elongate but lack cerci posteriorly. Diplurans are less common than the Protura but have the same habitat preferences. The Collembola, also known as 'springtails', are rounded or elongate in shape and generally shorter than 5 mm in length. They vary considerably in colour, and have a characteristic forked appendage (furca) which is used very effectively for jumping and is kept folded beneath the body when not in use. They feed on plant material and are commonly found in soil or leaf-litter and on plants.

CLASS INSECTA

ORDERS ARCHAEOGNATHA (Bristletails) and THYSANURA (Silverfish)
Vismotte, silwervissies

Often collectively known as the Thysanura, these insects are the most primitive in the class Insecta. They are elongate and wingless, and measure up to 20 mm in length. Projecting forwards from the head are long, multi-segmented antennae, similar in appearance to the paired cerci which are directed outwards from the posterior end of the body. A median caudal filament, often called the appendix dorsalis, lies between the cerci as a tail-like extension of the last abdominal segment. Vestigial limbs known as abdominal styles are sometimes present as small paired structures on the underside of the abdomen.

Of the two orders, the Archaeognatha is the more primitive. Its representatives have styles on most of the abdominal segments, and are further characterized by being more or less cylindrical, having large compound eyes, simple eyes, and a caudal filament which is often much longer than the adjacent cerci. The Thysanura closely resemble archaeognathans but are flattened, and have fairly small compound eyes and no simple eyes. The caudal filament is about the same length as the cerci, and there are few or no styles present on the underside of the abdomen.

Depending on the species, these insects may be active by day or by night. They tend to live in shaded leaf-litter, under logs, and on boulders and stones in shaded areas. Most species seem to be omnivorous and will eat a wide range of animal or plant matter. Several species are known to live in the nests of termites or ants, scavenging food from their hosts. The common nocturnal silverfish, *Ctenolepisma longicaudata*, is often found in houses in the KNP; it was introduced to South Africa from Europe through human trade practice.

The two sexes do not copulate directly, but have an elaborate courtship 'dance' during which the male gives off droplets containing sperm; the female then takes up these droplets by manoeuvring her abdomen over them. She deposits her fertilized eggs in a concealed place such as under a rock or log. The nymphs resemble the adults in body shape.

There is a school of evolutionary theorists which proposes that the ancestors of these two groups gave rise to the winged insects by growing lateral extensions on the second and third thoracic segments, possibly to facilitate escape by gliding away from danger.

Both orders are widely distributed throughout southern Africa but are not often seen because of their preference for living in concealed environments, and because they are often active only at night.

PLATE 1

ORDER EPHEMEROPTERA (Mayflies)
Eendagsvlieë

Adult mayflies are unlikely to be confused with any other insect: the forewings are large and triangular in shape, while the hind wings (absent in some species) are of the same shape but much smaller. Both pairs of wings usually have numerous crossveins. Projecting backwards from the abdomen is a pair of long cerci and, often, a central appendix dorsalis or caudal filament.

The nymphs of mayflies live in water, where they constitute an important item in the diet of many fish species. Most of these nymphs prefer cool, clear water conditions. Their shapes are reminiscent of those of the Archaeognatha and Thysanura (page 65), but they differ in having delicate protruberances arranged in pairs along the length of the abdomen. These protruberances are gills and are used for gas exchange, diffusing carbon dioxide out into the water and moving oxygen into the body. The nymphs generally feed on plant and animal material present in the water.

Because the nymphal stage is spent in water, the adults are generally found close to rivers and streams. Adults do not feed at all and have a very short lifespan, most species surviving for only a few days. On emerging from the subimaginal stage as adults they seek suitable mates and copulate, the females deposit their eggs in the water, and then they die.

Mayflies are fairly common throughout South Africa. They occur in and along all the rivers of the KNP. Adults are often attracted to lights at night during spring and summer.

PLATE 1

ORDER ODONATA (Dragonflies and damselflies)
Naaldekokers en waterjuffers

Similar in appearance to many species in the Neuroptera (page 91), the Odonata can nevertheless immediately be distinguished by their very thin, short antennae, and the fact that the wings are not held in a roof-like manner over the abdomen. The Odonata do not possess the wing-flexion mechanism which allows the wings to be folded back and held flat or roof-like over the body; instead, when resting on vegetation or elsewhere, dragonflies (Anisoptera) keep their wings extended sidewards much like those of an aeroplane, while damselflies (Zygoptera) hold their wings vertically above the body.

In both suborders, Anisoptera and Zygoptera, both pairs of wings are similar in shape and length, and are strengthened by an elaborate network of numerous crossveins. The wings of many species are ornately patterned or boldly coloured, and often have a metallic lustre. The bodies also tend to be very colourful but, unlike the wings, fade soon after death. This happens because the body colours are produced by epidermal pigments which denature after death, unlike the wing colours which result from light being reflected off the surface in a particular way, depending on the micro-texturing.

The adults are all predatory on other insects. They have the habit of perching on vegetation, or on soil or rocks in some cases, locating a suitably sized insect with their extraordinarily large eyes, then darting after the prey.

While most other insects mate by a temporary interlocking of the genital apparatus situated at the end of the abdomen, the Odonata have opted for a more circuitous and somewhat acrobatic method. Before mating, the male bends his abdomen forward in order to transfer sperm from the tip of the abdomen to a pouch under the third abdominal segment. This done, he flies off in search of a female. The male has claspers (modified cerci) at the posterior end of the abdomen and these he uses to hold the chosen female firmly behind her head. The female completes the fertilization by bending her abdomen forward until the tip of it touches the male's sperm pouch. The sperm is transferred while the pair is at rest, and they then fly around in tandem.

Dragonflies and damselflies are often seen skimming over a pool of fairly quiet water and then suddenly dipping down to lightly touch the surface before flying on: these are females laying eggs, which drift gently to the bottom of the pool and hatch after a few days.

The nymphs are aquatic and live in dams, streams or rivers, usually in the submerged vegetation along the fringes. In appearance, they vary from being robust and squat (dragonflies) to fairly slender (damselflies). All have well-developed legs and show the developing wings externally as flaps on the thorax. They prey on other small insects and water animals. To capture and hold their prey the nymphs have evolved a fascinating mechanism, found only in this order of insects. The lower lip of the mouth, called the labium, has been modified into a hinged, extensible arm with a scoop at its end. Normally held flat against the head, this structure is rapidly shot out to seize prey items swimming or drifting by.

For respiration, the nymphs of damselflies generally have three leaf-like gills projecting from the posterior end of the body; the nymphae of dragonflies, however, have an enlarged rectum the walls of which are lined with a network of tracheal branches (air tubes) which are continuously swathed with water so that gas exchange can take place. Additionally, in times of need, the water in the rectum can be forcefully expelled so that the insect is pushed forward, using the principle of jet propulsion.

Dragonflies and damselflies occur throughout Africa in most aquatic environments. In the KNP they are abundant along all large bodies of permanent water, although the adults often move far into the surrounding bush to search for prey.

PLATE 2 | **ORDER PHASMATODEA** (Stick insects)
Stokinsekte

Stick insects are usually medium to large in size, with some giants reaching up to 180 mm in length. They generally have an elongate, slender body and long legs. Winged species have the front wings narrowed and hardened, while the hind wings are large and soft, and are folded away beneath the front pair when not in use. The hind wings may be brightly coloured with eye-spots which the insect may suddenly flash open as a threat gesture when provoked. Such displays are nothing more than sham, however, and the insect relies for its safety on predators being frightened off by this behaviour.

The group contains some of the best examples of cryptic coloration in the animal kingdom. The various species specialize in blending into their environment, and in most this camouflage has reached a state of near-perfection. Many species have thorny outgrowths and knobby protruberances on their bodies, and veined, leaf-like wings which, together with their body colour, afford a resemblance so close to the surrounding vegetation that detection is extremely difficult. Their behaviour also aids in their camouflage: in a breeze they may sway gently from side to side, and they sit on trees and shrubs or assume such postures that their general shape and colour 'fits' the background.

Phasmids are phytophagous. In some parts of the world they may occasionally be found in great numbers when they cause severe damage as defoliators.

Many species of stick insects have very few male representatives, the vast majority of the population being made up of females; in some species no males have ever been found. The females of these species reproduce mainly by a process termed parthenogenesis, involving the laying of unfertilized eggs which give rise to viable and fertile offspring. This reproductive option is encountered in several of the insect groups. Females lay their eggs singly, usually on vegetation, or allow the eggs to drop to the ground, where they are often mistaken for plant seeds. The nymphs resemble the adults in general body shape and, in earlier instars, in addition to their camouflaging coloration have a remarkable defence strategy known as autotomy: when severely provoked or seized, many species allow a leg (or legs) to snap off at a special joint near the junction with the body. After escaping from the enemy, the insect regenerates a limb, to replace the lost one. Although often overlooked, stick insects are fairly common and widespread throughout the KNP.

PLATE 2 | **ORDER MANTODEA** (Praying mantids)
Hotnotsgotte

Often rather large and extraordinarily beautiful, praying mantids are easily recognizable by their elongate body shape, triangular head with its large compound eyes, and large raptorial forelegs with their tooth-like projections. In winged species the forewings are slightly hardened and serve as protective coverings for the large, soft hind wings. When threatened, praying mantids will sometimes flash open their hind wings in an attempt to frighten off the aggressor.

Praying mantids are active by day and are most often seen sitting on leaves, twigs or flowers. From these positions they slowly stalk insects which settle or walk nearby, lashing out with their scythe-like forelegs to grab hold of the prey.

The females of some species are cannibalistic and will seize and eat the male during or immediately after copulation. Although this habit may seem bizarre, it is the survival of the species that counts: by eating the male, the female is ensured of sufficient protein for her eggs to mature. The eggs are deposited on vegetation in a large, foamy ootheca, which quickly hardens. The ootheca is full of cavities, some of which contain eggs while others function as air spaces. The newly hatched nymphae resemble the adults except that they are wingless and smaller in size. With each moult the wing and body size increase, until at the final moult adult dimensions are attained.

Abundant throughout Africa wherever vegetation occurs, these insects are frequently seen in all areas of the KNP, but appear to be most common in shady, well-wooded areas adjoining the major rivers and streams.

PLATE 2

ORDER BLATTODEA (Cockroaches)
Kakkerlakke

Most cockroaches are broadly oval and flattened, with long, thin antennae and a head which is partially concealed beneath a large, shield-like pronotum. Projecting backwards from the abdomen are two short but squat cerci, highly sensitive appendages with their own direct nerve connections to the brain. The nymphae resemble the adults but for the wings, which increase in size at each moult, and sometimes they differ in colour. Some species remain wingless through the adult stage and are thus sometimes mistaken for immatures.

Because of the scavenging habits of a few species in domestic houses, cockroaches have earned a bad reputation and are generally looked on with revulsion. While there has never been any irrefutable evidence presented that cockroaches are a cause of human illness, they could conceivably promote illnesses such as salmonellosis (food poisoning) by walking over disease-infested filth and mechanically transporting it to human food.

Most species of cockroaches are active at night and spend their days in concealed places such as under logs and stones or resting quietly on a branch of a tree. The vast majority feed on organic offal, although some species have become specialist wood-feeders. These wood-eating species do not possess the enzymes necessary to break down the cellulose of plant cell walls, so they have established a symbiotic relationship with micro-organisms which do have this ability. The micro-organisms live in the intestines of the cockroach, processing the wood eaten by the host; they retain sufficient nutrients for their own well-being while the remainder are used by the cockroach.

Aggregating pheromones (secreted chemicals which serve to bring members of a species together) have been found in some cockroaches, while the males of others secrete a pheromone on to their abdomens which, when licked by a female, stimulates them to copulate. Females deposit their eggs in an ootheca. Depending on the species, the ootheca may be left in a concealed position, carried by the female as a large, leather-like pouch projecting from the end of her abdomen, or kept inside her body.

The cockroaches make up one of the oldest known insect orders, with fossil records which indicate that they have survived in an almost unchanged form for more than 300 million years. This evidence is a fair measure of their extraordinary success, and shows how flexible and effective they are in their ability to exploit a wide range of habitats and food sources.

Cockroaches are common throughout Africa. In the KNP they occur both in riverine forest and in the open veld. In caves frequented by large colonies of bats which provide a constant supply of nutrients through their faeces, enormous numbers of cockroaches of the genera *Gyna* and *Hebardina* are sometimes found; for example, more than 100 000 cockroaches frequent the relatively small interior of the Lanner Gorge Cave in the Pafuri area.

ORDER ISOPTERA (Termites)
Termiete

Often incorrectly referred to as 'white ants' or 'flying ants', termites are far removed from the true ants (Hymenoptera: Formicidae) in such aspects as evolutionary descent, body structure and habits. Like ants, however, termites are social insects and live together in colonies often numbering many thousands of individuals. The colony is divided into castes (workers, soldiers and reproductives), thus dividing the labour necessary for the continued existence of the colony. This division of labour results in a large saving of energy and is probably the main reason for the great success of social insects. Other social insects, such as ants and bees, have similar divisions and their success is equally obvious. Several families occur within the Isoptera, each having somewhat different habits and behaviour. Despite these differences they are similar in appearance and have similar life histories.

Termites have a poorly sclerotized cuticle (the skin is soft and membranous) and for this reason live in concealed or protected places such as inside logs and mounds (termitaria), or underground. Most termites feed on wood or other dead plant material. The only family to possess enzymes capable of breaking down the cellulose of plant cell walls is the Termitidae; the others rely on a symbiotic relationship with populations of single-celled organisms called flagellate protozoans which live in the intestines of the termites and do the digesting for them. The termites eat the wood and the flagellates break it down into components that they and the termites can assimilate. These intestinal micro-organisms live in the posterior region of the intestine, a region which embryologically forms part of the external skin or cuticle. This means that each time the skin is shed during normal growth the population of micro-organisms is lost. To renew the population of flagellates, the termite feeds on the faeces or an anal droplet of another termite which harbours these organisms.

The life cycle of the termite can be portrayed most easily by starting with the reproductives. Some time during spring or early summer masses of winged termites of the reproductive caste emerge from the nest. These mass emergences usually occur following periods of rain, so many colonies in an area may have reproductives emerging at the same time. The termites fly around for a short while. The females then settle on the ground or on a stalk of grass and each emits a pheromone, a faint scent which will eventually reach a male flying in the vicinity. The male follows the trail of the scent until he reaches the female. Something strange then occurs: the wings, which have been so essential for the dispersion of the individuals and the finding of mates, are discarded. To do this, the termite gives an almost imperceptible shrug, and the wings neatly snap off at a line of weakness near the point of attachment to the body.

Now wingless, the couple run off to find a suitable site for nesting. The female runs in front, the male following so closely behind that he continually bumps into her. A suitable site found, the couple make a hole in the soil or wood (depending on the species of termite), then copulate for the first time. These two termites will be the queen and king of the future colony. The queen will lay tens of thousands of eggs in her life, while the king survives to occasionally copulate with the queen and so assure her of an adequate supply of sperm.

Termites are heavily preyed on by other animals, in particular scaly anteaters, aardvarks, lizards, frogs, and other insects such as ants. Particularly in the Pafuri area, large numbers of termites are regularly killed by columns of foraging Matabele ants (*Megaponera foetens*).

As they generally live underground or in a concealed environment, termites are not often seen. Nevertheless they are very widespread and abundant throughout Africa, and the KNP is no exception.

There are differences in habits and appearance of genera within the Isoptera, and distinct termite groups (families and subfamilies) can be thus recognized. A brief discussion of some of the more common termites found in the KNP illustrates a few of these differences.

Family Hodotermitidae (Harvester termites)
Hodograsdraertermiete

An unusual feature of this family is that members of all castes have well-developed compound eyes, a characteristic possessed only by the reproductive castes in other termite families. Harvester termites can further be distinguished from members of other families in that they lack both a fontanelle and ocelli. The first segment behind the head, the pronotum, is fairly narrow and bent into a saddle-like shape.

Harvester termites have a preference for fairly dry grassland areas. No above-ground mounds are constructed; rather, the nest consists of a collection of interlinked tunnels and chambers below the soil, with a number of inconspicuous openings to the exterior. Some of the chambers are used for day-to-day activities such as grooming, passing of food and breeding, while others are used as storage areas for food. During times when there are no foraging trips or other external activities in progress, the holes linking the nest with the outside world are sealed off with moist soil.

Harvester termites feed on plant material, generally dead grass but occasionally also green parts of living plants. This vegetation is collected by the workers which forage during the cooler parts of the day. The food is carried back to the nest and stored in chambers for later consumption. Micro-organisms living in the intestines of the termites break down the plant material into digestable units (page 70).

Only two species of this family occur in southern Africa, and of these only one, *Hodotermes mossambicus*, is found in the KNP. This species is widely distributed throughout most of the drier areas in subSaharan Africa, and in some areas is a serious competitor with vertebrate herbivores for available grass.

Family Kalotermitidae (Dry-wood termites)
Droëhouttermiete

Termites of this relatively primitive group can be recognized by their lack of a fontanelle, by the ocelli usually present on the upper part of the head, and by their having the pronotum flat and wider than the head. The family has no true worker caste; the immature reproductives perform all the tasks done by the workers in other families.

Unlike other termites, species in this family make nests within logs, or a dead branch still attached to a tree. They generally bore through the outer layer of the branch, then construct the nest by tunnelling and excavating occasional chambers. The chewed wood is used as food; none appears to be stored for later use (storage of food is unnecessary as these insects live within their food supply). Dry-wood termites have a preference for soft wood, this being easier to chew and in which to construct nests. The termites depend on symbiotic micro-organisms living in their intestines to digest the wood for them (page 70).

PLATE 3

Family Termitidae

Most termite species existing today occur within this family, which can be further divided into four subfamilies, three of which occur in the KNP. Compared with species in the other termite families, all these subfamilies are advanced in an evolutionary sense. The Termitidae produce their own enzymes capable of digesting the cellulose of plant cell walls, which has freed them from dependence on intestinal micro-organisms (page 70). Some species make use of other organisms to aid in digestion, such as the Macrotermitinae, which use fungi in the breakdown of plant material. A wide range of habitats, habits and body structures is displayed by the various species: some live underground and some in mounds; some feed on grass, others on wood, and some specialize in cultivating and eating fungi.

The subfamily Termitinae includes wood- and humus-feeding species which live secluded and secretive lives below the soil or in mounds. Only on rare occasions do any of the termites venture into the outside world. The workers tunnel below the soil and consume any dead and rotting wood in their path. They then return to the nest, where they feed hungry members of the colony by regurgitating or excreting a drop of partially digested food. Some species of Termitinae form exceedingly large colonies which may number many thousands of individuals.

The subfamily Macrotermitinae contains the fungus-feeding termites. Like the close association formed between intestinal micro-organisms and termites of other families (page 70), so the Macrotermitinae have established a symbiotic relationship with certain fungi. Termite and fungus derive mutual benefit from this association, so much so that the one cannot exist without the other.

The very large termitaria often seen in the north of the KNP are constructed by members of this subfamily. Several mounds which can be seen from the road along the Luvuvhu River at Pafuri measure well over four metres in height and seven metres in circumference; more than a million termites may be living within such a mound. Inside the termitaria the termites live in a manner which is both extraordinary and fascinating. In fact, it is difficult to comprehend the faultless organization and functioning of such a large society, a seemingly mindless conglomerate of individuals functioning almost as loosely adhered organs within and for the survival of a single body.

Situated in a central chamber deep within the mound are the king and queen. The king is an average-sized termite, but the queen may reach a length of up to 100 mm. The bulk of her body consists of a grossly swollen abdomen which is filled with egg-producing organs. The queen is constantly attended by workers which clean, feed and protect her, and take care of the eggs. Spread around the royal chamber, within which the king and queen are permanently cloistered, are scattered passages and air spaces where the other termites live, and which by their structure ensure that the temperature and humidity remain relatively constant throughout the day.

Near the royal chamber in carefully built pockets lie a series of fungus gardens. Workers collect dead plant material outside the nest and chew, swallow, semi-digest and excrete it. The faeces is then moulded into a sponge-like comb on which the fungus is actively cultivated. The small, white reproductive bodies that eventually appear abundantly on the fungus are fed to the king, queen and young termites. The soldiers and workers feed on the comb itself, after it has been reduced by the fungus to components which the termites can assimilate.

The 'snouted termites' belong to the subfamily Nasutitermitinae. Their common name derives from the unusual appearance of the soldiers, which have very small mandibles, but have the front part of the head drawn out into a pointed, hollow snout through which a repellant substance is sprayed at enemies from a large gland in the head. Together with the other castes of the colony, which have normal heads and functional jaws, these snouted termites live in small to medium-sized mounds which have a rounded or domed appearance. The colony feeds almost exclusively on grass collected by workers which go on foraging trips at night. In places where these termites are abundant, large areas of grass may be denuded, and in many parts of Africa they are serious competitors of livestock for this resource.

ORDER ZORAPTERA

Few people, including entomologists, are likely to notice this small order of obscure insects. They are elongate, less than 3 mm in length, generally pale, and have well-developed legs and chewing mouthparts. Both winged and wingless forms exist. Most have a strong avoidance reaction to light, and are usually found in decaying wood or humus. They appear to feed on fungal spores, mites and other small organisms.

Although representatives of this group have been collected from various widely distributed locations in Africa and elsewhere, little is known of the true extent of their distribution or of their biology in general. They have not been found in the KNP.

ORDER ORTHOPTERA (SALTATORIA) (Grasshoppers, locusts, crickets and their relatives)
Sprinkane, krieke

This large order embraces many widespread and common species. Most members have well-developed legs with the hind pair enlarged and adapted for jumping. The body is generally elongate and medium to large in size.

Most species are winged, with the forewings hardened as a leathery covering to protect the soft, membranous hind wings when at rest. Many species have special structures to produce sound as a method of attracting mating partners, and these species also have highly developed auditory organs to receive these sounds.

Many species found in Africa (although not necessarily in the KNP) and other parts of the world are of great economic importance, such as the desert locust, *Schistocerca gregaria*, the red locust, *Cyrtacanthacris* (*Nomadacris*) *septemfasciata*, and the brown locust, *Locustana pardalina*. These are the locusts that form the migratory swarms, often spreading across many kilometres, which wreak havoc on crops in their path, sometimes resulting in human starvation and great economic loss. Nevertheless, Orthopterons form a vital component of our ecosystem, particularly as a food source for smaller vertebrates such as frogs, lizards, snakes, birds and shrews.

The Orthoptera is divided into two suborders, based on distinct differences in habits and appearance.

SUBORDER CAELIFERA

This very large group comprises all the grasshoppers and locusts, as well as some lesser-known forms. Members of this suborder generally have short antennae with less than 30 segments.

In those species which stridulate or 'sing', sound is produced by the insect rubbing its hind legs against its forewings or against the side of its abdomen, or by the hind wings rubbing against the forewings. The auditory organ is situated on the first abdominal segment, diagonally above the hind pair of legs.

The vast majority of insects in this suborder are active by day.

PLATE 3 **Family Acrididae** (Grasshoppers and locusts)
Korthoringsprinkane

This is the most abundant and widespread group in the Orthoptera, with numerous species in Africa and on other continents. There is a great variation in shape, size and coloration among the different species. Many are cryptically coloured, often a brilliant green to conceal themselves among foliage or else a dull, broken patchwork of different shades of brown and black, making them difficult to detect on rocks, soil or dry vegetation.

Most acridids are active by day and feed on plant material. Although a few species lay their eggs in decaying wood, females generally deposit their eggs below the soil, using the ovipositor to laboriously dig a sufficiently deep hole. After emerging, the nymphs, which resemble the adults in shape, crawl to the surface and commence feeding. With each moult their wings increase in length, until adult structure and size are attained.

Adults are heavily preyed on by small mammals, birds and lizards, while the eggs are sought after by the larval stages of certain beetles and other insects. The lifeless bodies of grasshoppers which have been parasitized and finally killed by fungi are also frequently seen suspended on vegetation.

Acridids are common throughout most areas of Africa. In the KNP they can be found on vegetation, stones and soil in dry, open bushveld or lush, riverine forests. The red locust, *Cyrtacanthacris septemfasciata*, occurs endemically in the KNP and is especially common on the northern plains. Here it exists in stable populations not known to form swarms. The traditional outbreak area for red locust swarms is in the periodically flooded, grass-filled depressions in parts of Tanzania, Zambia, Malawi and Mozambique. From here enormous swarms migrate into invasion areas which include South Africa. Such large-scale invasions into the KNP were recorded in 1906, the 1930s and 1940s.

A bigger economic problem in South Africa is the brown locust, *Locustana pardalina*, but this species does not occur endemically in the Transvaal. A large invasion by brown locusts into the Kruger National Park was recorded during the mid-1920s by Colonel James Stevenson-Hamilton.

PLATE 3

Family Pyrgomorphidae

Although pyrgomorphids have a typical grasshopper or locust build and shape, many species in this family are striking due to their vivid contrasting coloration. Colours often present are yellow, red and black, and serve to warn potential predators that these species are distasteful and, often, poisonous. Young and inexperienced predators may attempt eating these insects, but soon learn to associate the colours with distastefulness and so to avoid them. The members of this family are generally sluggish in their movements, without the well-developed jumping abilities found in locusts and grasshoppers of the Acrididae.

The phenomenon of using bright colours to warn predators of distasteful tissues and secretions, called aposematic coloration, is common to many insects. Many harmless species mimic distasteful species by copying their colours and body shape, and are thus also avoided by predators; this is known as Batesian mimicry. Where two totally different but equally distasteful species use the same warning colours, this has the effect of strengthening the avoidance reaction in predators; this is called Müllerian mimicry. The large *Phymateus* individuals occasionally seen during the warmer months have the habit of spreading and rustling their wings as a defence mechanism, frightening off would-be predators.

Pyrgomorphids are fairly common in the drier areas of Africa but only a few species occur in the KNP. They are generally seen feeding on herbaceous plants. *Zonocerus elegans* is a fairly common resident in the KNP, and aggregating groups of hoppers of this species are often seen in spring and summer. As they grow they adopt more solitary habits and are then generally found as individuals.

SUBORDER ENSIFERA

Members of this group may be recognized by their antennae, which have 30 or more segments and are longer than the body. Many species are able to produce sound – such as the familiar noise of the garden cricket – by rubbing together the two forewings, which have specially modified and strengthened veins at the base. Those species with hearing organs have them situated on the forelegs, in many cases appearing as a slit in a slight depression just below the 'knee' (the femoro-tibial joint).

PLATE 4 | **Family Tettigoniidae** (Katydids)
Langhoringsprinkane

The Tettigoniidae form a large group and are divided into 19 sub-families. Feeding habits vary considerably, some species feeding on plant material, some being carnivorous and others omnivorous. Most species are winged and live in trees or shrubs, while the wingless species prefer vegetation closer to the ground. Females often have a very long, curved, sickle-like ovipositor and generally deposit their eggs in slits cut in plant tissue, although a few species lay their eggs in soil.

Also called 'long-horned grasshoppers', most katydids are strikingly beautiful insects, often seen sitting on flowers or leaves. Many have a delicate green colour with fragile, multi-veined hind wings which are kept carefully tucked below the leathery, leaf-like forewings. However, forming part of this family are the rather grotesque but quite harmless armoured ground crickets or 'koringkrieke' (subfamily Hetrodinae). Sharp, spiky outgrowths protrude from parts of the body, enhancing their fierce appearance. They generally occur in dry areas and may be seen walking on the ground or sitting on branchlets of shrubs or low trees. Although they are plant feeders, they may occasionally scavenge on dead insects, including members of their own species which, during periodic population explosions, are killed in fairly large numbers on roads. In this species the hind wings are absent, and the forewings in the males have been reduced to short stubs which are vigorously rubbed over each other to produce a loud rasping sound to attract a mate; the females are wingless and mute.

Members of this family occur in most of the well-wooded parts of Africa. In the KNP they are most abundant in the lush vegetation along rivers and streams.

PLATE 4 | **Family Gryllidae** (Field crickets, tree crickets and their relatives)
Veldkrieke

Members of the Gryllidae have three-segmented tarsae. The females of most species have a long, thin, straight ovipositor and deposit their eggs in soil, either singly or in groups, but some of the subfamily Oecanthinae lay theirs in slits deep inside twigs. Many species live below ground, while others hide under rocks, logs or leaves by day and forage at night. Most are omnivorous, feeding on plant or animal matter as the opportunity arises.

Included in this family are the common garden crickets which dig holes and tunnel under our lawns. Besides the medium-sized, blackish *Gryllus bimaculatus*, which is common in lawns in the KNP, some light-brown individuals, *Brachytrupes membranaceus*, measuring more than 70 mm long, have been found at Pafuri and Skukuza. These unusually large crickets make tunnels below the soil, generally 150-250 mm deep, with a chamber at the end. Besides being the largest, this species is reputed to be the loudest of all crickets in the world. Also forming part of this family are the ghost or tree crickets (subfamily Oecanthinae), pale, creamy-white, elongate insects which are occasionally found sitting on broad-leafed shrubs, mostly along rivers and streams or in densely wooded areas. They chew pear-shaped holes in the leaves, then partially protrude their body through one of the openings and raise their forewings to seal the gaps. When sound is produced the leaf acts as a 'baffle' which increases the intensity of the sound. Only males produce sound in this way, attracting females over considerable distances.

The family Gryllidae is widely distributed throughout Africa and is abundant in the KNP, especially in well-wooded areas near water.

PLATE 4

Family Gryllotalpidae (Mole crickets)
Molkrieke

The mole crickets are all specialized burrowers in soil, this being particularly obvious when examining their strong forelegs, which are shortened, broadened and flattened to increase their effectiveness for subterranean tunnelling. They are active at night and are often attracted to light. They do not bite, but when held in the hand they attempt to burrow between the fingers with their powerful forelegs, which may create the sensation of being bitten or gnawed.

Mole crickets generally feed on the roots of plants, but occasionally may also feed on dead insects or fleshy remnants.

Four species are known to occur in southern Africa, *Gryllotalpa africana* being the most common. In the KNP this species is plentiful in damp areas well covered with vegetation, such as along river banks. They are not often seen, however, because of their subterranean and nocturnal habits.

PLATE 5

ORDER DERMAPTERA (Earwigs)
Oorkruipers

Elongate insects with medium-sized legs and antennae, earwigs are generally coloured in shades of brown, black or red. Because of their shortened forewings and general body shape, earwigs superficially resemble staphylinid beetles (page 98), but most species may be distinguished by the modified cerci, which look like a pair of stout pincers or forceps, at the posterior end of the body. Also characteristic is the form of the wings in those species capable of flight: the forewings are very small, hardened, and almost square in shape while the hind wings are large and soft, roughly semicircular, and when not in use are pleated and doubled back so that they are folded below the forewings.

Most earwigs are active at night and have the undeserved reputation of crawling into the ears of humans, something which has never been proved to happen but which has undoubtedly resulted in innumerable of these insects being killed.

When threatened or cornered, earwigs often raise the tip of the abdomen, displaying the cerci in a threatening gesture, but they are harmless. They do not inject poison through the cerci and the effect of their grip on a human finger or other part of the body is negligible. The modified cerci are used for threat displays in defence, in mating, and possibly in the procurement of food.

There are two suborders of earwigs in Africa. The Forficulina, which are by far the most common, are scavengers on plant and animal matter, or prey on small insects. They are generally nocturnal, but are often attracted to lights. By day they retreat to dark, protected locations such as beneath logs or stones, and under fallen leaves or rotting vegetation. They are found throughout the KNP in most habitat types.

The suborder Hemimerina (Diploglossata) contains nine or 10 unusual species which are semi-parasitic on rats of the genus *Cricetomys*, and live in the fur and feeding on flaking skin, hair debris and a fungus which grows on the skin of the rat. Generally about 10 mm long when fully grown, these species are orange-brown in colour, have no eyes or wings, and are capable of rapid movement to escape grooming behaviour. The cerci are not modified into pincer-like structures as in the Forficulina, instead being fairly straight, hairy, rod-like appendages. All members of this suborder belong to the genus *Hemimerus* and have thus far been collected in Africa only between 10 °N and 20 °S. They do occur in Zimbabwe and the northern Transvaal, but have not been found in the KNP.

PLATE 5

ORDER EMBIIDINA (Web-spinners)
Tonnelspinners

Web-spinners are rarely seen because of their inconspicuous body form, dull coloration and secretive habits. Small to medium-sized insects, they are uniformly coloured in shades of brown, having a slightly flattened, elongate body. The legs are short and stout, and a pair of two-segmented, squat cerci project posteriorly from the abdomen. They are easily recognized by the swollen or enlarged appearance of the segment near the end of the foreleg (the proximal segment of the tarsus), caused by glands from which the silk used for the construction of their nests is derived. The males have soft, membranous wings which are used for dispersal flights to enter other colonies for mating and gene exchange.

Feeding mainly on plant material, web-spinners live in small colonies within nests or galleries under stones or logs, below bark or in a similar protected environment.

Web-spinners are distributed throughout the warmer areas of Africa, although in the KNP they have been collected only at Pafuri and Skukuza. They are probably fairly common but remain unnoticed because of their concealed way of life.

PLATE 5

ORDER PLECOPTERA (Stoneflies)
Pêrelvlieë

The adults of this order found in the KNP are elongate, medium-sized insects with a dull brown body colour. They have fairly long, multi-segmented antennae and two pairs of multi-veined, membranous wings. The forewings are not hardened to form protective coverings for the hind wings. These insects are weak fliers. The legs are well developed, as are the two antenna-like cerci projecting from the posterior end of the body.

After mating, the female deposits her eggs in a river or stream. The eggs float to the bottom and attach themselves to the substrate by means of adhesive papillae. Whereas the adults live on land, the nymphae are aquatic. They resemble the adults in general body shape except in lacking wings, and having slightly longer cerci and external gills in many species. Nymphs of the Plecoptera are most easily distinguished from the nymphs of the Ephemeroptera (page 66) by having only two appendages, the cerci, at the hind end of the body. The nymphs are most frequently found under stones or on stone-littered sand in fairly fast-flowing water. Here they prey on other insects and small animals such as tadpoles. Nymphae may live for as long as two years and moult up to 30 times.

When fully grown the nymph climbs on to an object projecting from the water's surface, and the adult emerges through a split which develops along the upper surface of the larval body. The adults are not, as their common name implies, restricted to living on stones, but are found near rivers, resting on shrubs, trees, stones and logs. Unlike the nymphs, adults are short-lived, having a life span of a few days.

Although widely distributed throughout Africa, stoneflies of the family Perlidae do not appear to be abundant anywhere. In the KNP only this family is represented, adults of which have been collected only along the Sabie River near Skukuza and the Luvuvhu River at Pafuri. However, this is probably because collecting efforts have been poor, as they are almost certain to occur in and along all the major rivers and streams. At Pafuri they appear to be fairly common during summer, frequently being attracted to lights near the Luvuvhu River at night.

The Nemouridae is the only other family which occurs in Africa, and is restricted to the southern regions. The adults of this family have silvery-grey wings and the nymphs live in clear, cold mountain streams, mainly in the southern and south-western Cape, but also in the northern and north-eastern Transvaal. Both adults and larvae feed on plant material.

PLATE 5

ORDER PSOCOPTERA (Booklice or psocids)
Boekluise

Although this is a fairly large order and contains several species of economic importance, few people take much notice of these rather secretive insects. They tend to be shorter than 5 mm in length, are elongate but robust, have long, slender antennae and chewing mouthparts, and very often have a bulging post-clypeus. Winged and wingless species occur; in some species only the females have wings.

Psocids are common in most environments and may be found on vegetation, under bark, in stored thatching and grain, in birds' nests, or even in the plumage of birds or on the bodies of animals such as elephants. Many species living on the bark of trees construct a silken canopy for protection.

Although most are gregarious and scavenge on plant and animal matter, several species of *Liposcelis* occur in or on books where they may feed on the paste of the book-bindings or mould on the paper, a habit which has given rise to the misleading name 'booklice'.

Psocids are widely distributed on most continents. They are common throughout the KNP.

PLATE 6

ORDER PHTHIRAPTERA (Lice)
Luise

Most members of this very widespread order tend to be rather flat in appearance, only a few millimetres in length, and have poorly developed eyes and short, stubby antennae which are sometimes hidden in grooves on the head. None has wings or cerci. Their body form has adapted to their parasitic way of life: their small size and flattened shape enable them to remain close to the body surface of the host and so provide less of a target for the bird or mammal to locate, catch or scratch off; and many species have the distal ends of their legs modified so that they resemble pincers, ideally shaped to grasp and hold on to feathers or fur.

All lice derive their nutrients from birds or mammals. They are dependent on their hosts for food during all stages of the life cycle, and remain permanently on these hosts. Lice generally are able to disseminate when host members make close physical contact, by climbing on to other hosts living in the same nest, roost or burrow, or even by clinging to a biting fly and dropping off when the fly alights on a new host.

The eggs are attached to the feathers or hair of the host. Immature stages are similar in appearance to the adult stage. Nymphae undergo three moults before reaching adulthood.

There are four suborders of lice, each having its own characteristic adaptations and way of life.

SUBORDER AMBLYCERA

These are biting lice with short, four- or five-segmented antennae which are generally concealed in grooves on the side of the head. The mandibles move in a horizontal plane, so that the insect bites with a side-to-side action. These lice feed on a wide range of substrates, including feathers, skin debris, serum, blood, and even their own eggs and nymphae.

The largest members of the Phthiraptera are found in this suborder, some species measuring 12 mm in length. In the KNP *Laemobothrion maximum* has been collected from martial eagles; other genera found include *Colpocephalum*, *Hohorstiella*, *Meromenopon*, *Numidicola*, *Myrsidea*, *Dennyus* and *Actorniphilus*.

SUBORDER ISCHNOCERA

These biting lice have fairly long antennae which are not concealed in grooves on the head. Their mandibles are arranged in a vertical plane, so that the insect bites in an up-and-down manner. This suborder differs from the Amblycera also in that the third antennal segment is not pedunculate (stalked).

This suborder contains only two families, the Philopteridae which in the Afrotropical region occurs on birds, and the Trichodectidae, which occurs on mammals. They appear to feed on shreds or remnants of host skin, feathers and hairs, serum and sebaceous secretions.

Seven genera of the Philopteridae have so far been collected from birds in the KNP; of the Trichodectidae two common genera are *Damalinia* and *Trichodectes*.

PLATE 6

SUBORDER RHYNCHOPHTHIRINA

This small and unusual suborder contains only two species. In both, the front part of the head is prolonged into a slender, downcurved rostrum, with the mandibles placed at the very tip. The mandibles are unusual in that they have rotated through 180 degrees and thus work outwards; they are thought to be used to rasp away at the skin of the host in order to draw blood, which the insect then drinks.

Haematomyzus hopkinsi lives on warthogs and has been collected only in Kenya and Uganda. The other species, *H. elephantis*, has been

found parasitizing the African elephant, from the Sudan to the Transvaal, and has also been found on the Indian elephant.
H. elephantis is widespread on elephant thoughout the KNP. These lice are not easy to locate as they blend well with the skin colour of the host and generally occur in very low numbers.

PLATE 6

SUBORDER ANOPLURA

These are the 'sucking lice', so called because their mouthparts resemble a syringe mechanism and are used to puncture the skin of the host, enter a blood vessel and suck out blood. These mouthparts have been modified into a piercing structure, retractable into a sac-like extension of the head when not being used. The Anoplura occur only on mammals, and include the three species which parasitize humans, namely *Pthirus pubis*, *Pediculus humanus* and *Pediculus capitis*.

Pthirus pubis, as its name suggests, lives in the pubic hairs of humans, but may spread to other parts of the body where coarse hair is present (the eyebrows and axillary hair). They are about 2 mm long, have a crab-like appearance and are pale grey in colour. They cement their eggs on to the pubic hairs; the larvae, when they hatch, often remain unnoticed as they do not cause much irritation. The adults, however, are capable of producing severe irritation due to their biting into the skin for blood. Surprisingly little is known of the medical importance of this species under natural conditions, but it is not generally considered to be important as a vector of diseases.

In contrast, *Pediculus humanus* has been responsible for great suffering and the loss of millions of lives because of the diseases it transmits. Generally known as the body louse, it concentrates in areas which are constricted by clothing, such as the armpits, waist and shoulders. Bites from these lice may result in irritation and an inflammatory reaction. More importantly, however, are the diseases they may transmit, which include epidemic typhus (also called louse-borne typhus). The causative micro-organisms of this disease, *Rickettsia prowazeki*, are ingested by lice when feeding on an infected person. These then multiply in the gut of the louse and are passed out with the faeces. Humans are infected with the disease when the faeces are rubbed into cuts or abrasions on the skin, or even when dried faeces are inhaled or come into contact with the eyes. Also carried by *Pediculus humanus*, and caused by *Rickettsia quintana*, is murine typhus; this disease is normally transmitted by fleas but the body louse does occasionally become a carrier. Epidemic relapsing fever, caused by *Borrelia recurrentis*, is also transmitted by *P. humanus* and infects humans when spirochaetes (a type of bacterium) in the haemolymph of crushed lice enter the bloodstream through abraded skin.

The head louse, *P. capitis*, is restricted to the hair on the head, and is not considered an important vector of human diseases.

These three species of human lice have an almost cosmopolitan distribution, being most abundant in places where humans live under filthy and crowded conditions. They have occasionally been collected from humans in the KNP, but no disease resulting from their presence has been recorded in the area. Members of the genera *Linognathus* and *Haematopinus* are common on several of the larger species of mammals in the KNP. The genus *Polyplax* contains many small species commonly found on mice and other small mammals.

ORDER HEMIPTERA (Bugs)
Eesies

The fifth largest order of insects, this group includes pests such as aphids, scale insects and stink bugs, as well as species of medical importance such as bed bugs and assassin bugs.

Hemipterans have a characteristic, elongate, sucking proboscis. Although this is not nearly as long as those found in the moths and butterflies it can nevertheless be seen as an extension of the mouthparts into a distinct rostrum. Depending on the species of bug, the rostrum is inserted into plant tissues for the ingestion of nutritive sap, or into mammals or other insects for blood or body fluids.

There are two suborders. These differ so distinctly that they are sometimes considered separate orders.

SUBORDER HOMOPTERA

The majority of species in this group are winged, although in certain species both sexes may be wingless; in some species only males are winged. The wings are uniformly membranous and not divided into different regions as in the Heteroptera (page 86), and are held in an angled, roof-like manner over the body.

The Homoptera receive their nourishment from juices present in plants. Using their mouthparts they pierce leaves, soft stems or roots and then imbibe the sap.

PLATE 7

Family Cicadidae (Cicadas or Christmas beetles)
Sonbesies

Robust hemipterans often measuring well over 20 mm in length, cicadas have a broad head and a stout thorax and abdomen, the latter being covered by the well-developed wings when the insect is at rest. Although some species are attractively coloured, most are rather drably patterned in shades of grey or brown.

Cicadas are well known for the incessant, loud, buzzing noise they make during the hot summer months. This is produced only by the males, but both sexes have hearing organs for the reception of sound. The sound-producing and hearing organs are situated on the abdomen, at its junction with the thorax, slightly behind the last pair of legs. Muscles vibrate a membrane to produce sound, which is magnified by air sacs. The sound has a ventriloquial quality, which explains why it is often difficult to pinpoint a cicada by sound alone.

Cicadas are generally found sitting on the bark of trees, were they are well camouflaged and difficult to locate. The adults live for a few weeks only, their sole purpose being to find mates and copulate so as to ensure the following generation.

The female cicadas lay their eggs in twigs or branches of trees and shrubs. When the nymphs emerge from the eggs they drop to the ground and proceed to dig into the soil, where they may live for many years; the forelegs are thickened and strong, and well adapted for digging. The nymphs feed on plant juices by inserting their tube-like mouthparts into the roots of trees.

Fully grown nymphs time their emergence from the soil to coincide with the beginning of the rainy season. At this time, they make their way to the surface and climb up a tree or shrub until well above ground. Here, the skin splits along the upper part of the body and the winged adult crawls through the split.

Cicadas are widely distributed and common in the warmer parts of Africa. As the nymphs spend their time below soil only the adults are normally seen, and then only for a few weeks during the rainy season, generally between October and January. Cicadas have been found throughout the KNP.

PLATE 7

Family Fulgoridae (Fulgorids or lantern flies)
Lanternbesies

Fulgorids tend to be medium-sized insects, generally 10 to 15 mm in length. Many species have the anterior part of the head projected forwards as a tube-like growth; this feature presumably serves to break up the general body shape and in so doing it enables the insect to escape notice by predators. Most fulgorids are drably coloured, which allows them to blend well with the twigs or bark on which they rest. Some species, however, are brightly coloured in brilliant light-green.

A curious feature associated with the nymphs and adults of many species of the Fulgoridae is the fan of delicate, white, thin, wax filaments which spreads out from the hind end of the body. The function of this fan has yet to be determined.

Fulgorids occur in all the warmer, well-vegetated parts of Africa. In the KNP they are most often found resting on twigs and branches of shrubs and trees in areas of luxuriant vegetation.

PLATE 7

Family Cercopidae (Spittle bugs)
Skuimbesies

The Cercopidae are generally small insects, 10 mm or less in length. They are notable for the exceptionally good jumping ability of the adults and the unusual protective coverings the immature stages construct for themselves.

Both adults and young are sap feeders on plants. The young find themselves a suitable spot on a soft branch or grass stem, push their long mouthparts into the sap-conducting vessels of the plant, and then surround themselves with a frothy, spittle-like mass which is produced by a gland and accessory apparatus located near the anus. The froth protects their soft, immature bodies from the hot sun, and probably also from predatory insects.

In some species the production of froth is so copious that it literally drips down from the tree, which has given rise to the common name 'rain-tree' for some tree species in which this insect lives. Some cercopids have nymphs which live underground on the roots of plants, and even here a surrounding mass of spittle is produced.

Cercopids are common throughout Africa and occur in all parts of the KNP. The adults are generally found resting on vegetation.

PLATE 7

Family Cicadellidae (Leafhoppers)
Blaarspringers

Leafhoppers are common but because of their size and coloration they are easily overlooked. The adults are generally less than 10 mm in length, and many species are only 2 or 3 mm long when fully grown. They are very similar to cercopids (page 83) in shape. Most cicadellids are light green or brownish, although some are striped and brightly coloured.

Leafhoppers are responsible for the transmission of viral diseases from plant to plant, as they suck the sap of an infected plant and then pass on the virus to other plants on which they feed.

Cicadellids are common throughout Africa. They have been collected in all areas of the KNP.

PLATE 8

Family Membracidae (Treehoppers)
Boomspringers

The Membracidae is a family of small, mostly squat insects. Many species have extraordinary growths protruding from the pronotum; these projections of the cuticle extend upwards and backwards over the body, and occasionally form near-uncanny structures. Because many treehoppers are dark in colour, these growths give them the appearance of a piece of bark, a thorn or a bird dropping and thus protect them from predation. Other species have the projections shaped into twin, symmetrical forms.

Like many other homopterons, treehoppers exude a sweet substance, known as honeydew, from the posterior end of the body. Ants are known to feed on this excess sap, and for this reason are often seen grouped around a feeding membracid, avidly lapping up its exudations. Large numbers of treehoppers may occur on individual trees, and in such cases their excessive feeding activities frequently inhibit the growth of the tree.

This family is widespread in all the warmer parts of Africa and member species occur throughout the KNP. They are most often found sitting on smaller branches of trees and shrubs.

PLATE 8

Family Psyllidae (Psyllids)

Small insects resembling miniature cicadas (page 82), psyllids are most often found in groups on leaves or on the soft, thin branches of shrubs. The nymphs are flat and oval in shape, whereas the adults are more like cicadas and have well-developed jumping and flying capabilities.

As sap feeders, psyllids may be responsible for the transmission of several plant diseases. Additionally, certain psyllids have an ingredient in their saliva which causes depressions or galls to form on the leaves at their feeding sites. Psyllids often occur as pests on citrus trees throughout Africa.

In the KNP they are most frequently found in areas of lush vegetation such as along river banks. During spring and summer massive outbreaks may occur in the northern mopane veld, when literally hundreds of thousands of small, green, adult *Arytaina mopani* feed on mopane leaves and they are attracted in huge numbers to lights in the camps at night.

Immatures of this species, which resemble scale insects, also occur on mopane leaves. The immatures are hidden beneath a creamy to honey-coloured waxy scale, and their feeding activity causes the mopane leaves to curl and twist so that the insects are further protected against both the harsh sun and potential predators.

PLATE 8 | **Family Aphididae** (Aphids or plantlice)
Plantluise

Small insects measuring only a few millimetres in length, aphids tend to be barrel-shaped and most are green in colour. They are very common and are generally found gathered on plant stems, leaves and flowers. They feed by inserting their mouthparts into the sap-conducting vessels of a plant. The pressure in these vessels is so strong that the sap simply flows into the aphid without the aphid having to suck at all. Aphids are notorious for the large number of diseases they may transmit from one plant to another.

The life history of many species is complex and interesting. Within a single species a season may start with a winged male and female reproducing sexually, and the female laying eggs. The offspring may all be wingless when they reach the adult stage, the females of which may then reproduce parthenogenetically, giving birth to a series of young without first having had to mate. These offspring may in turn give rise to a series of wingless, parthenogenetic generations, after which winged offspring may again appear, reproduce sexually, and start the whole cycle over.

A fascinating association sometimes occurs between aphids and ants. The sap on which aphids feed is high in sugar content but has relatively few proteins. To gain sufficient proteins, therefore, the aphid takes in large quantities of sap, most of which it excretes after having extracted the protein but not all the sugar. Ants, notorious for their attraction to sweet liquids, have developed the habit of gently prodding an aphid with their antennae. At this signal the aphid excretes a droplet of sugar-laden sap, which is consumed by the soliciting ant. Certain ant species may protect aphids from parasites and predators, and may even move them to shelter during winter in areas with harsh climates.

Aphids occur throughout Africa wherever plants grow. In the KNP they are most common in areas of lush vegetation, but they also live on plants in dry, open bushveld areas.

PLATE 9 | **Families Diaspididae and Coccidae** (Scale insects)
Dopluise

The Diaspididae and Coccidae are sap feeders on plants. They spend the greater part of their lives living beneath a protective, waxy shield which they secrete, the shield increasing in size as the insect grows. The families differ in the toughness of the shield: Diaspidids are generally known as armoured scale insects, whereas Coccidae are referred to as soft scale insects.

The first-stage nymphs, or 'crawlers', are mobile and have legs, antennae and eyes. They find a suitable spot on a plant, settle down, insert their very long mouthparts into the plant, and moult into the second stage. During this moult they lose their legs, eyes and antennae, and become essentially a sac with a tube into the plant. The protective shield appears soon after the first moult. They go through two (in females) or three (in males) further moults, after which the adult emerges. The females are wingless and remain permanently under the shield they secrete. Males emerge with eyes, legs and a single pair of wings, looking very much like tiny flies. They fly off, find themselves a female, mate and die.

Scale insects are found all over Africa wherever plants occur. Several species are pests on fruit trees. They have been found throughout the KNP in open bushveld and shaded riverine forests. Although frequently found on stems and leaves of shrubs and trees, and even on grasses, members of the Diaspididae are perhaps most obvious on succulent, fleshy leaves of the aloes in many rest camps in the KNP. Here they may be seen as grey or whitish, waxy spots dotted in large groups over the surface, and to non-entomologists not even by the wildest stretch of imagination could they be described as insects. If one of these 'dots' were lifted, however, a soft and delicate little animal – although one that still does not resemble a familiar insect – would be found below.

SUBORDER HETEROPTERA

Like the Homoptera, many heteropterans are plant feeders, but there are also a large number of species which prey on other insects. Whereas homopterans have the first pair of wings uniformly textured, winged members of the Heteroptera usually have the first pair divided into a hardened basal part and a soft, membranous apical region. The wings are held flat over the body and not in an inverted V-shape as in the Homoptera.

PLATE 9 | ### Family Cydnidae (Burrowing bugs, stink bugs)
Graafwantse

These insects are generally 4 to 6 mm in length, dark in colour, flat and oval, and have very stout spines along their legs. Perhaps they are best known for the strong, unpleasant smell they emit when disturbed or stepped on, a characteristic which has given rise to their alternative common name, stink bugs.

Burrowing bugs live in soil or beneath leaf-litter and stones. Large numbers are occasionally attracted to lights on warm summer evenings.

Although not often seen, burrowing bugs are common in the warmer, well-vegetated areas of Africa, and are distributed in large numbers throughout the KNP.

PLATE 10 | ### Family Pentatomidae (Shield bugs)
Skildstinkbesies

The most commonly seen shield bugs are about 10 mm long, and have a flattened body which is roughly triangular in shape. They vary considerably in colour, some being drab while others are bright and beautiful. If disturbed they often emit an unpleasant-smelling secretion from glands which open on the thorax.

Most shield bugs are sap feeders on plants, although some species are predatory on other insects.

Shield bugs are usually seen as solitary individuals, in most vegetated areas of Africa. They are fairly common and widespread in the KNP.

PLATE 9 | ### Family Coreidae (Coreids, squash bugs)
Blaarpootwantse

Coreids resemble assassin bugs (page 87), except for the large, almost triangular prothorax and the large number of veins on the membranous end-part of the forewings. They often have very stout, enlarged hind legs, which sometimes have projections on them. Most are medium-sized. In colour they tend to be rather drab, either brown or charcoal-black. Like shield bugs and burrowing bugs (see above), many of these insects are capable of giving off a powerful, repellant smell if disturbed.

All are plant feeders, and occasionally may be seen in groups feeding on the sap of shrubs or other plants.

Coreids are widely distributed and common in all the warm, vegetated areas of Africa. Many species occur abundantly in the KNP.

PLATE 9
PLATE 10

Families Lygaeidae and Pyrrhocoridae (Seed bugs, cotton stainers)
Grysstinkbesies, rooiwantse

The Lygaeidae, a large family, and the Pyrrhocoridae, a small family, are discussed together as they are so alike in both appearance and habits. Most of these insects vary between 5 and 15 mm in length. The most commonly seen species are brightly coloured with red and black markings on an orange background. Their bodies are elongate, and they are somewhat reminiscent of assassin bugs (see below).

Most feed on plant sap or seeds, and some species are pests on fruit trees and shrubs. Both families contain 'cotton stainers', species which feed on the bolls of cotton plants, leaving holes through which fungi enter and discolour the cotton.

They are well represented in the warmer, vegetated areas of Africa, and have been collected from most areas of the KNP.

PLATE 10

Family Reduviidae (Assassin bugs)
Roofwantse, sluipmoordwantse

These commonly seen insects gained the name 'assassin bugs' because of their habit of stalking other insects, on which they prey. They are generally medium-sized, although many small and large species do occur, and have a great range of body colours. Like praying mantids, many species have raptorial forelegs to aid in grasping prey. It is not advisable to handle assassin bugs, as many species do not hesitate to bite and this may be very painful.

Ectrichodia crux is a large, orange and black species which is specialized to feed on millipedes. These bugs have modified front legs to enable them to hold on to the smooth cuticle of the millipede. Groups of them have been found preying on millipedes at Skukuza and Pafuri.

The Reduviidae is a large, common and well distributed family, occurring throughout Africa, including in the KNP.

PLATE 10

Families Cimicidae and Polyctenidae (Bed bugs and bat bugs)
Weeluise

The Cimicidae are generally flattened, oval or elongate in shape, and have reduced forewings resembling small flaps. They have compound eyes and three-segmented antennae.

All five of the nymphal stages and the adults survive on blood which they suck from humans, birds and bats. Only two species of the family Cimicidae are parasites on humans, however, these being *Cimex lectularius*, which is distributed widely throughout the world, and *C. hemipterus*, which is restricted to the tropical regions. Normally hidden in crevices, cracks or loose material, they come out at night and crawl on to the host to feed.

Although they have been found under natural conditions to be infected with a wide range of disease organisms – including *Wuchereria*, *Brugia*, *Trypanosoma* and *Brucella* – cimicids have never been proved to actually transmit any disease.

Several species of cimicid bugs have been found parasitizing bats in the KNP, including at least two species which are new to science. Members of the genus *Cacodmus* appear to be the most common and are most frequently found on bats of the genera *Pipistrellus*, *Eptesicus* and, occasionally, *Scotophilus*.

The Polyctenidae are also flattened and have greatly reduced forewings. They differ from cimicids in having no eyes, four-segmented antennae, and short, flat spines arranged in one or more rows behind the head. All polyctenids are parasitic on bats, and spend most of their lives on the host.

Polyctenid bugs are viviparous, having only two larval stages comprising the life cycle. No member of this family has been collected in the KNP, but they have been found parasitizing *Rhinolophus fumigatus* elsewhere in the Transvaal.

PLATE 10	**Family Miridae** (Plant bugs)
	Plantwantse

Although they are common, plant bugs generally are not noticed because of their small size; they usually measure less than 10 mm in length. Most are sap feeders on plants, although some are predatory on small arthropods.

Plant bugs make up a very large family and occur throughout the vegetated areas of Africa.

PLATE 11	**Family Gerridae** (Water striders)
	Waterlopers

The Gerridae is a semi-aquatic family of medium-sized insects having long middle and hind legs. The legs and ventral surface of the body have a densely packed layer of water-repellant hairs which prevents them from sinking.

The insects of this family either scavenge, or prey on other insects which have fallen into the water from the river bank or overhanging trees. They pierce their prey with their pointed mouthparts to extract the insides.

Water striders are generally seen skimming along the surface of slow-moving water at the edges of rivers, or in dams or temporary pools. They are well distributed in suitable habitats over most of Africa, and have been recorded throughout the KNP wherever suitable bodies of water occur.

PLATE 11	**Family Notonectidae** (Backswimmers)
	Rugswemmers

As their name indicates, backswimmers have the unusual habit of swimming upside-down. Medium to small in size, they swim by means of enlarged hind legs which are flattened like oars and are equipped with fringes of hair. The hairs increase the surface area for 'pushing' during the propulsion stroke, but then fold back with the recovery stroke to lessen friction.

By using their hind legs, backswimmers can move about very rapidly, swimming just below the water's surface, darting down to the murky depths, and searching out and catching the insects on which they prey.

These insects have a longitudinal keel stretching along the midventral line of the abdomen. On each side of the keel is a hair-covered groove where air is stored, enabling the insect to breathe below water.

Backswimmers are found in quiet pools or in slow-moving water at the edges of rivers. They are widespread and common in Africa, and are found throughout the KNP in suitable bodies of water.

PLATE 11 — **Family Corixidae** (Water-boatmen)
Bootmannetjies

Yet another family of insects commonly found in quiet water near the edges of rivers, water-boatmen superficially resemble backswimmers. They are elongate and have a pair of large, oar-like hind legs which are fringed with hairs. Water-boatmen are easily distinguished from backswimmers, however, as they do not swim upside-down, and are flatter in appearance.

Water-boatmen spend most of their time at the bottom of pools but come to the surface occasionally to replenish their supply of fresh air. Many species are predatory on other aquatic insects or small water organisms, but most feed on microscopic algae.

Water-boatmen are common and widespread throughout much of Africa including the KNP.

PLATE 11 — **Family Nepidae** (Water-scorpions)
Waterskerpioene

Daunting in appearance, water-scorpions may grow to a large size. They are generally brown in colour. They derive their common name from the enlarged front legs which resemble the pincers of true scorpions, and the long posterior siphon resembling a sting. Like praying mantids, they use their front legs to grasp and hold the prey which they hunt among roots and plant debris of shallow river margins or ponds. Prey items include tadpoles or suitably-sized aquatic insects.

The 'tail' has no sting and is actually a respiratory siphon (breathing tube) with openings at the end. By holding this 'tail' up to the surface for oxygen the insect can remain below the water, searching for prey. It is best to avoid these insects as they may give a painful bite if handled.

Water-scorpions are fairly common in rivers and large pools. They occur throughout the warmer parts of Africa, including the KNP, where they are usually found as solitary individuals.

PLATE 12 — **Family Belostomatidae** (Giant water bugs)
Reusewaterwantse

In this family we find the giants among the hemipterans. The species most frequently seen in the KNP is a large, broad, flat insect which is brown to charcoal-black in colour and has stout legs.

Giant water bugs are similar in habits to water-scorpions (see above).

Both water-scorpions and giant water bugs prey on insects, tadpoles or even small fish which they find in submerged vegetation or between rocks on the bottom of rivers or large pools. A space between the top of the abdomen and the wings is used for air storage, enabling the insect to breathe when below the water.

Belostomatids are occasionally attracted to lights at camps which are situated near rivers, where they announce their arrival with a loud whirring sound and a sudden thud as they land ungracefully against a window or wall. Like water-scorpions, they are capable of inflicting a very painful bite if handled.

Giant water bugs are distributed over much of the warmer areas of Africa. They have been collected from all the major rivers in the KNP.

PLATE 12

ORDER THYSANOPTERA (Thrips)
Blaaspootjies

Because of their small size (they seldom grow to longer than 6 mm) and inconspicuous habits, thrips are not often seen despite their relative abundance. They have an elongate, cigar-shaped body and well-developed compound eyes which resemble bunches of grapes, each eye being composed of a number of easily distinguished facets. Each leg ends in an inflatable vesicle (sac) which can be everted or retracted by pumping haemolymph in or out. This enables the insect to walk on a wide range of substrates. The winged species are most easily recognized by their two pairs of feathery wings, each wing being made up of a narrow shaft fringed with numerous setae; not all species have wings, however.

The mouthparts are modified into a piercing or rasping and sucking organ by asymmetrical adaptation of the maxillae and mandibles to form three stylets within a tube-like structure. The stylets are prodded into plant tissues for feeding on sap, or into the bodies of other insects, depending on whether the species is a plant feeder or predacious. Many thrips feed on fungal spores.

While certain plants depend on the Thysanoptera to pollinate their flowers, several species of this order are injurious to plants. Some cause unsightly blemishes on fruit when they feed in the axil region of the fruit-bud; as the fruit grows larger these feeding sites expand as scar tissue. Others damage the leaves or flowers, some cause the formation of galls, and a few have been found to transmit diseases from plant to plant.

The Thysanoptera have two larval stages in their breeding cycle, and a 'resting stage' between the larval and adult stages, which corresponds to the true pupal stage of the more advanced orders of insects. This 'resting stage' is actually a period of intense cellular activity during which the insect undergoes the drastic metamorphosis from larval to adult stage, a metamorphosis that generally results in major changes in appearance and habits.

The larvae resemble the adults in general appearance, except in that they lack wings; this stage is followed by a prepupal stage, generally of a short duration, and finally the pupal stage (or 'resting stage'), which is characterized by a lack of mobility and the presence of wing pads, occurs.

Thrips are widespread throughout most areas of Africa including the KNP, and occur in leaf-litter and in the flowers of many plants.

PLATE 12

ORDER MEGALOPTERA (Alderflies)
Bergstroomjuffers

Two families in this order occur in Africa, and both are restricted to mountainous areas in the southern part of the continent. The Sialidae is found in the Cape, while the Corydalidae has been recorded from the Cape and Natal. Superficially resembling Neuroptera (see below), these medium-sized to large insects differ in having the hind pair of wings broadened near the base, and when at rest fold these broadened areas (anal area) fanwise. All are predacious on other insects or small aquatic organisms.

Corydalids deposit large numbers of eggs in compact batches near streams, and the hatched larvae make their way to the water where they usually live below rocks in fast-flowing areas. Gas exchange takes place through the eight pairs of filamentous gills which are arranged on either side of the body. The larvae are also capable of surviving for lengthy periods in dried-up stream beds. Pupation occurs in the soil adjoining the stream or in dense vegetation on the soil.

Alderflies have not been recorded from the KNP.

ORDER NEUROPTERA (Lacewings, antlions and their relatives)
Netvlerkinsekte

Many of the more common species of Neuroptera are often mistaken for Odonata (page 66), which they superficially resemble. Like the Odonata, most Neuroptera adults likely to be seen in the KNP have an elongate, rounded abdomen, and two pairs of elongate, multi-veined wings. The Neuroptera can be distinguished as they fold their wings back, roof-like, over the body; the Odonata cannot do this as they have not evolved the necessary hinge mechanism. The antennae of the Neuroptera may be short or long in the adults but are always conspicuous, in contrast to those of the Odonata, which are very thin, short and difficult to see.

Most adults and all larvae are predatory, usually on soft-bodied insects. Most species spend their entire lives on land, but a few have larvae which live in water. The larvae have their mouthparts modified into hollow, piercing structures, often sickle-shaped, which are jabbed into the prey to suck out the body fluids.

PLATE 13

Families Hemerobiidae and Chrysopidae (Lacewings)
Netvlerkies

Fairly common, especially at outside lights on warm summer evenings, these medium-sized insects tend to display the typical Neuropteran body shape and form. Adult hemerobiids are generally dull brown in colour, whereas chrysopids are usually a delightful, bright green.

Unlike hemerobiids which attach their eggs directly to leaves or bark, female chrysopids suspend their eggs on stalks so that ants and other predators cannot reach them. To do this, the female touches the substrate (generally the bark of a tree) with the hind end of her abdomen, giving off a droplet of clear, viscous liquid. She then raises her abdomen, drawing the droplet into a thin thread which soon hardens, and lays the egg at the end of this stalk. The larvae of both families are free-living, generally on vegetation, where they prey on aphids and other small insects. The larvae are elongate, with a slightly bulging or oval abdomen.

Chrysopids have been collected throughout the KNP. Hemerobiids appear to be somewhat rarer, although they are also widely distributed in most areas of the KNP.

PLATE 13

Family Mantispidae (Mantispids)
Valshotnotsgotte

Mantispids offer an excellent example of convergent evolution, where two unrelated groups of organisms have independently evolved the same structure for the same purpose. Mantispids have their forelegs modified in the same way as have praying mantids (page 68), to serve as raptorial or grasping structures for catching and holding prey. The resemblance does not end here, however: the prothorax in mantispids has been extended into an elongate neck, as it has in the praying mantids, thus increasing the manoeuvrability and effectiveness of the enlarged forelegs. Mantispids are easily distinguished from praying mantids, however, by having both pairs of wings soft and membranous, the front pair not being hardened, and both pairs being of similar size and shape.

Adults are generally seen sitting on shrubs and flowers, patiently waiting for a prey item to move into range. The larvae of some species are parasitic on spiders, while other species have larvae which seem to be predatory on beetle larvae living in soil.

The females lay their eggs on stalks in the same way as those of the Chrysopidae (page 91). After hatching, each larva (of the type that are parasitic on spiders) runs around in search of a suitable spider's-egg cocoon, tunnelling its way into the cocoon once it has been found. The larva waits within the cocoon until the spider's eggs have hatched, then feeds on the young spiders by piercing their bodies with its tube-like mouthparts and drinking the body fluids. When it is fully grown the larva spins itself a cocoon inside the empty spider's-egg cocoon, then pupates. The adult gnaws its way through these cocoons to the outside world.

Mantispids have been recorded from many parts of Africa, but appear to be fairly rare and not often seen. In the KNP they have been recorded from several localities.

PLATE 13

Family Myrmeleontidae (Antlions)
Mierleeus

This is by far the most common of the Neuropteran families occurring in the KNP. The medium to large adults resemble dragonflies (page 66), but their cumbersome flight easily distinguishes them from the latter, as do their prominent antennae and the manner in which their wings at rest are folded back, roof-like, over the abdomen.

Myrmeleontids are perhaps best known for the characteristic, conical pits that are made by the larvae of most species. They construct these by moving backwards in a circle, digging into the soil as they go and flicking away excess soil with their heads. Once the pit has been dug the larva lies in wait at the bottom, covered by soil, until a suitable insect stumbles into the pit. The prey insect struggles to escape as the walls of the pit cave in and the antlion flicks sand at it. The antlion then embeds its large mandibles into the insect and begins feeding. Large numbers of these antlion pits sometimes may be found within a relatively small area.

When fully grown, the larva digs into the soil and spins itself a protective silken cocoon. Sand grains adhere to the silk, which is sticky when first secreted, further protecting the larva by forming an encrusted layer around the cocoon. Adults are nocturnal; by day they are most often found sitting on vegetation or leaf-litter in shady, wooded areas, especially riverine forest.

Myrmeleontids are common in most parts of Africa. In the KNP they are most often found in the vicinity of rivers and streams, although they do occur elsewhere.

PLATE 14

Family Ascalaphidae (Owl flies)
Langhoringmierleeus

Owl flies closely resemble adult myrmeleontids in general body form, but have long antennae which are strongly clubbed at the end. Adult ascalaphids may be recognized by their habit of raising their abdomen at right angles to the body when resting on vegetation, often giving them the appearance of a broken twig.

The females lay their eggs in clumps on grass, other vegetation or rocks. The larvae are very similar in appearance to the larvae of the Myrmeleontidae, but do not dig pits in the soil. Instead, they roam above ground among vegetation or below rocks and logs, hunting for small insects on which to feed. Most adults appear to be active in the evening, preying on other insects by catching them in flight.

Ascalaphids are widely distributed throughout Africa. In the KNP they frequent riverine forest and other wooded areas, but are not very common and are not often seen.

PLATE 14

ORDER SIPHONAPTERA (Fleas)
Vlooie

Despite their small size, fleas are easily recognized by their distinctive appearance. The body is strongly flattened laterally, and is most often orange-brown to dark brown in colour. The hind legs are well developed and much enlarged, and are used by the insect to jump considerable distances. Projecting backwards from the prothorax is a row of stout bristles (in many species) which, together with other bristles on the body, assist in hooking the insect into the hairs or feathers of the host and thus reduce the risk of its being dislodged. The mouthparts are modified in adults into a piercing proboscis for sucking blood.

Females lay their rather large, whitish eggs either on the host or on the floor of the host's nest, burrow or dwelling. The larvae are legless and elongate, and feed on organic material from the host, such as flaking skin. They also feed on undigested or partially digested blood defecated by the adults but, with the exception of one species from Australia, the larvae are never directly parasitic. When fully grown the larvae pupate within a cocoon in a crack or crevice, or among debris. They may remain in the cocoon for several months, not emerging until a host is nearby. They detect the presence of a host by vibrations and, often, by an increase in the level of carbon dioxide. They then jump on to the host and feed by piercing the skin and sucking blood.

All adult fleas feed on blood by parasitizing birds and mammals. Each species of flea generally has a narrow range of preferred host species, but sometimes they feed on other species as well. Only those birds and mammals which have nests, dens or burrows, or at least frequently return to the same place, tend to be consistent hosts of fleas. Because the supply of blood from the host is abundant, fleas are not concerned with extracting all the available nutrients from every mouthful of blood: they feed voraciously, extracting those nutrients which are easily removed and excreting the remainder as partially digested blood.

Fleas feeding on a species of mammal or bird may occasionally jump on to another species of mammal or bird when these are in close proximity, such as from a rat on to a human. Because of this habit, and their habit of feeding on blood, fleas are known to transmit disease, not only between members of the same species but also between members of different species.

One of the oldest and historically most devastating of human diseases is that known as the Plague which, during the Middle Ages, caused the deaths of countless millions of humans and is estimated to have killed up to half the population of England and one third of the population of Europe during the 'black death' of the mid-fourteenth century. Plague, caused by bacteria called *Yersinia pestis*, is primarily a disease of rodents and is endemic in many parts of the world. The main reservoirs of the disease in southern Africa are gerbils belonging to the genera *Tatera* and *Desmodillus*. The multimammate mouse, *Mastomys coucha*, which often flourishes around human habitation, is important in that it carries plague from these wild reservoirs to domestic situations.

Plague is generally transferred from rodents to man by fleas of the genus *Xenopsylla*, although it may also be transmitted by other fleas, including *Nosopsyllus fasciatus*, *Leptopsyllus segnis*, the human flea, *Pulex irritans*, and several others. Fleas feeding on rodents infected with plague take up the bacteria with the blood. The bacteria then multiply so rapidly in the gut of the flea that they form a plug. When the flea jumps on to a human and attempts to feed, the plug of bacteria may become dislodged and the disease organisms may be regurgitated into the bloodstream of the human host. Plague can then be transferred from one human host to another by fleas normally associated with man or, more rarely, as air-borne bacteria from sneezing and coughing (pneumonic plague). Although there have been no major outbreaks of plague in South Africa, many people have died, usually in the rural districts, during localized, small outbreaks.

Another disease normally associated with rat populations is endemic murine typhus (*not* epidemic typhus or typhoid fever). This is transmitted from rat to rat, or rat to man, by fleas.

The human flea, *Pulex irritans*, the dog flea, *Ctenocephalides canis*, and the cat flea, *C. felis*, are all intermediate hosts of the dog tapeworm, *Dipylidium caninum*, which may establish itself in and parasitize children when they accidentally ingest infected fleas while playing with household pets.

The female of the sticktight flea, *Echidnophaga* spp., partially burrows into the skin of the host, often causing an ulcer to form around her. She then lays her eggs in this lesion, after which most of the eggs eventually drop to the ground; those that remain in the lesion don't hatch. The hatched larvae feed on organic debris in the soil. Of the six species present in Africa the best known is *Echidnophaga gallinacea*, which is occasionally found clustered in a dense mass around the heads of domestic fowl in tropical areas. In the KNP, *E. larina* has been found parasitizing warthogs, bushpigs and the African elephant, while *E. aethiops* has been found on the ears of *Nycteris thebaica* bats in the Pafuri area.

Known as the chigoe, chigger, jigger or sand flea, *Tunga penetrans* was introduced into the tropical zones of Africa from tropical America. After copulation, females burrow into the skin of the host (often man, but it also parasitizes several other mammals), usually that of the feet. The females are particularly fond of digging into the soft skin between the toes or under the toenails of a human host. When completely burrowed in only the posterior extremity of the female is visible, and she drops her eggs to the ground from there. The female swells to the size of a small pea; to relieve the pain the host may scratch or squeeze the offending female, which often results in her bursting. Inflammation and gangrene may set in, sometimes resulting in the affected limb requiring amputation. This species does not occur in the KNP.

ORDER COLEOPTERA (Beetles)
Kewers

The Coleoptera is the largest of the four major insect orders, the others being the Diptera, the Lepidoptera and the Hymenoptera. The Coleoptera consists of more than 300 000 described species, and includes the smallest living insect (which is much smaller than a pinhead) and the bulkiest (a scarabaeid longer than 150 mm), and the largest family in the animal kingdom (the Curculionidae).

The word 'coleoptera' means 'sheath-winged', and describes the nature of the insects in this order. With a few exceptions, adults have two pairs of wings, the front pair hardened into strong, protective covers called elytra. The elytra are always shorter than the soft, membranous, hind wings. Because of their greater length, the hind wings – which do the actual flying – are elaborately twisted and folded each time they are tucked below the elytra. Most adult beetles are protected by a heavy, well-developed cuticle. Besides affording excellent protection from predators, this thick cuticle ensures a minimal loss of water through the skin by evaporation, a major contribution to the phenomenal success of beetles as terrestrial insects.

The beetles present a bewildering array of life styles, encompassing predators, parasites, scavengers and plant feeders. Some live in water, others in trees, many on land, and some in the bodies of other insects. While most beetles have a dark coloration, many are brilliantly coloured.

The larvae do not have the protective outer skin as well developed as the adults, and most have a grub-like form. They are found burrowing in soil, trees or animal and plant debris, depending on the food preference of the species.

Of the Coleoptera families present in the KNP, many are rare and not likely to be seen. Only the more frequently encountered families, those in the suborders Adephaga and Polyphaga, are discussed here.

SUBORDER ADEPHAGA

Beetles in this suborder may be recognized by the coxae of the hind pair of legs, which are fused to the thorax and are so arranged that the first abdominal segment appears divided. A further characteristic is a groove, or suture, where the dorsal cuticle of the pronotum joins with the propleuron.

PLATE 15

Family Carabidae (Ground beetles)
Grondkewers

Generally dark in colour, these beetles vary considerably in size, although most are medium to large. The vast majority have large mandibles to catch and hold prey effectively. Most adults depend on their well-developed legs to run down their victims on open ground, so that many species have lost the ability to fly, their elytra having become fused together and the hind wings reduced or absent.

Carabids are generally seen as individuals on the ground, although many also live under bark or on vegetation. Most of them are active by day. Most adults are predators on other insects in the larval and adult stages; a few species are phytophagous, feeding on seeds or other parts of plants; and some species are general scavengers.

One of the more conspicuous species to be found in the KNP belongs to the genus *Anthia*. Occasionally seen on open ground, it is a large (sometimes reaching 50 mm in length) black beetle with a yellow mark on either side of its pronotum. It has long, curving mandibles and an elongate body. If threatened or attacked it can squirt from its anus a repellent stream of formic acid for a distance of about 350 mm. This substance causes a burning sensation on human skin, and is extremely painful if squirted into the eye.

Carabids are very common throughout Africa. They have been recorded in all the vegetation zones of the KNP.

PLATE 15

Family Cicindelidae (Tiger beetles)
Sandkewers

These may well be termed the supreme athletes of the beetle world. Their prowess in running and flying is far greater than that of most other beetles. This rapid mobility, combined with their keen eyesight and large mandibles, makes them very efficient as predators on other insects. They tend to be medium-sized, have an elongate body that rests on long, agile legs, and often have brightly patterned elytra. Most are active by day, although some are nocturnal (it is in this latter group that wingless species occur).

The larvae are equally efficient as predators. They have the grub-like form of most beetle larvae, but with a well-sclerotized head capsule and a hump on the abdomen. They construct vertical tunnels in the soil, from which they ambush unsuspecting insect passersby. The hump on the abdomen has a few hard hooks which are pressed into the sides of the tunnel to prevent the grub from slipping down.

Tiger beetles are widespread and common in the warmer areas of Africa. In the KNP they are most often seen on sandy banks next to rivers, sometimes in fairly large numbers. The largest members of this family in the KNP, dwarfing most tiger beetles found anywhere in the world, are those of the genus *Mantichora*. They may measure as much as 50 mm in length and males have enormous curved mandibles. These magnificent flightless beetles are most frequently seen in the Pumbe sandveld region north of Nwanetsi.

PLATE 15

Family Paussidae (Paussids)

Paussids tend to be fairly small insects, most being less than 10 mm in length. They attract attention because of their curious, club-like antennae, but are seldom seen as they live in association with ants in their nests. The ants feed on secretions given off by the beetle, and the beetle in turn apparently feeds on the ants.

As is the case with some species in the Carabidae (page 96), some paussids are able to squirt a caustic fluid to deter potential predators. Occasional specimens are attracted to lights at night.

Paussids have been found in ants' nests in many places in Africa. In the KNP they have been found only in the extreme north and south, but this is probably a reflection of inadequate collecting efforts.

PLATE 15

Family Dytiscidae (Diving beetles)
Waterkewers

Aquatic insects, the adults are flattened and streamlined, both adaptations for living in water. The hind legs are much stouter than the other legs, and are modified as highly efficient oars. By using these flattened legs, fringed with hairs, the insect can propel itself forward at great speed. Close examination of the forelegs of males will often reveal that the distal part is flattened and bears a number of suction pads. These pads enable the male to cling to the smooth body of the female during copulation.

The adults are air-breathing and carry their supply of air in a space between the wings and the abdomen; the spiracles open into this space so the insect can breathe when below water, and needs to surface only once in a while to replenish the supply. The larvae, which are elongate and also live in water, have breathing pores at the posterior end of the body. When in need of fresh air they have to return to the surface.

Adults and larvae are predacious and live on other insects or small animals such as tadpoles. The larvae are equipped with large mandibles which operate like a syringe: the prey item is held firmly with the mandibles while digestive enzymes are injected into the victim. This reduces the insides of the prey to a nutrient-rich broth, which is then sucked up by the larva.

Diving beetles prefer slow-moving water or quiet pools, and are found throughout Africa in most places where such conditions exist. They have been found in all the rivers of the KNP, as well as in many of the temporary pools which form after rain.

PLATE 15

Family Gyrinidae (Whirligig beetles)
Waterhondjies

Adult whirligig beetles are generally oval in shape and approximately 10 mm in length. They are streamlined and have the middle and hind legs flattened to serve as 'oars'. Their long front legs are used to capture prey.

When examined closely, whirligig beetles appear to have four eyes, two on either side of the head; in fact, each eye is divided in two and separated so that the insect can see above and below the water simultaneously. Adults are found swimming on the surface of the water in groups, unlike diving beetles which tend to be solitary and spend most of their time below water.

Both adults and larvae are predatory, feeding on small animals which share their aquatic environment. The larvae are elongate in shape and have tufted gills projecting from the abdominal segments.

Whirligig beetles are common in slow-moving waterbodies throughout most of Africa. They occur in all permanently flowing rivers of the KNP.

SUBORDER POLYPHAGA

The vast majority of beetles belong in this group. All differ from the suborder Adephaga (page 95) in that the ventral surface of the first abdominal segment is continuous, not divided into two separate sections by the coxae of the hind legs. There is no visible groove or line indicating the junction of the pronotum with the propleuron. Despite these unifying characteristics, the beetles in this suborder vary widely in appearance and habits.

PLATE 16

Family Histeridae (Histerids)
Harlekynkewer

Histerids are often seen at animal dung, but are especially abundant at carcasses in the earlier stages of decomposition. Those found in the KNP vary from about 1 mm to 15 mm in length.

The most commonly seen histerids are those of the genus *Saprinus*. These are generally less than 10 mm in length, have short legs, and a body that is rounded, smooth, hard, and metallic-blue in colour. The genus *Hister* is also fairly common, with several species occurring in the KNP. These may be up to 15 mm in length, shiny black in colour and have striations on their elytra. Both *Saprinus* and *Hister* are voracious predators of dipterous (fly) larvae found in dung or carrion. Several other genera exist but are less conspicuous in both appearance and habits. Some are greatly flattened and live under bark, others inhabit the tunnels of woodboring insects, while many species occur in the nests of termites and ants.

The larvae are grub-like with short legs but have well-developed, large mandibles. They are predatory on other insects and often occur in the same locations as the adults.

As a family, the Histeridae is widely distributed in Africa. The genera *Saprinus* and *Hister* have been found throughout the KNP.

PLATE 16

Family Staphylinidae (Rove beetles)
Dwaalkewers

Rove beetles are easily recognized by their narrow, elongate shape and greatly shortened elytra, which generally cover only about half the abdomen. They have fully developed, large hind wings which are elaborately twisted and folded beneath the elytra when not in use. These beetles are proficient flyers and runners. There is a great range in size within the Staphylinidae; species collected in the KNP range from 2 mm to 20 mm in length. The most commonly seen species are about midway in this range of measurements, and dark in colour, although some of the larger species may be brightly coloured.

As is true with most large families, there is considerable variation in habits and habitat within this group. Most rove beetles are predatory on small animals such as other insects; some are scavengers, others plant-feeders, while some live with termites and ants in their nests. The genus *Aleochara* has species which are parasitic on the pupae of cyclorrhaphous diptera.

Several thousand species of rove beetles occur throughout Africa. In the KNP they can be found in soil or leaf-litter, running about on land, in dung, at carrion, or on shrubs and trees.

PLATE 16

Family Lycidae (Net-winged beetles)
Platvlerkkewers

Frequently seen sitting at the tip of a stout grass stem in open veld, these are flattened, medium-sized insects, often with expanded or broadened, waxy forewings indented with striations. Most are orange in colour, with a black band or marking at the posterior end of the forewings. The bold colours warn potential predators that the insect is distasteful, and the same colour pattern is mimicked by certain long-horn beetles (page 105) and moths (page 124).

Net-winged beetles are found mainly in the tropical regions of Africa. They are fairly common in the more open bushveld areas of the KNP, but little is known about their general biology.

PLATE 16

Family Lampyridae (Fireflies)
Glimwurms

These medium-sized insects often attract attention at night by their flashing light signals, which serve to attract a mate. All species have at least one sex capable of producing this luminescence, and the various species have different lighting sequences so that only members of the same species are attracted to each other. The females of some species mimic the signals of other species, and when the male arrives the female devours him.

All adult males are winged but the females of some species retain the larval form, which is elongate and multi-segmented, resembling a series of flat plates. These wingless females are often called 'glow-worms'. Males are often seen in groups, flying slowly while signalling. The larvae of most species feed on snails and slugs, injecting digestive enzymes through their grooved mandibles into the prey, then drinking up the resulting broth. The adults are reputed not to feed at all.

The light-producing organs – situated ventrally at the posterior end of the abdomen – consist of light-emitting cells overlying cells containing crystals which reflect the light outwards. Emission of light results when the enzyme luciferase interacts with a substance called luciferin within the light-emitting cells.

Found throughout the warmer bushveld areas of Africa, fireflies and glow-worms are widespread in the KNP, and are most common during summer.

PLATE 17

Family Cleridae (Checkered beetles)
Bontroofkewers

Clerid beetles are generally attractively coloured, elongate in shape, and vary in length from approximately 6 mm to 15 mm. Most species are predatory on woodboring beetles in trees and shrubs.

The bright blue or blue-green *Necrobia rufipes* is common at carrion, where it feeds on flesh and fatty deposits, and preys on small blowfly maggots. The fully grown larva of this species has been found to crawl into an empty blowfly puparium, close the entrance with a white silky mass, and pupate there.

Most commonly found in the tropical areas of Africa, clerids are widespread in the KNP, with *Necrobia rufipes* being the most often seen.

PLATE 17

Family Lymexylidae

These are thin, elongate insects, generally 30 mm to 40 mm in length, and most having shortened elytra. The hind wings are unusual in that they are not tucked under the forewings when at rest, but are held fanwise, in a rather untidy manner, over the abdomen. Little is known of the habits and general biology of these insects, except that the larvae of some species tunnel in hard wood.

There are approximately 50 species in the family, many of which occur in Africa. Several specimens have been collected in the KNP, all after having been attracted to lights in the evening.

PLATE 17

Family Elateridae (Click beetles)
Kniptorre

Adult click beetles range in size from fairly small to rather large, and are generally oblong in shape and dull brown or grey-black in colour. They often have backward-directed projections on either side of the prothorax. Some species are predatory, while others feed on plant material. The larvae, which are elongate in shape with a well-sclerotized, yellow-brown cuticle, are often found tunnelling in rotten trees or soil.

A characteristic feature, and an effective defence mechanism used by the insect, is the flicking motion produced when the tension of two body parts pressed against each other is suddenly released. This is accompanied by a dull 'click' sound. The flicking motion and the sound result when a tough, elongate outgrowth below the prothorax slips forcibly into a groove ventrally on the mesosternum. The insect resorts to this action when turned on to its back or held by a predator; when holding one of these beetles, its action may have a disconcerting effect as it appears that a finger or part of the hand may be caught between the pro- and mesothorax, with the result that the beetle is dropped and can escape.

Click beetles are fairly common in the warmer and well-wooded areas of Africa. In the KNP they occur in riverine forests and in the drier bushveld zones.

PLATE 18

Family Buprestidae (Metallic woodborers, jewel beetles)
Pragkewers

Most of the adults in this family are medium-sized insects, and are easily recognizable by their brilliant metallic colours and compact, bullet-shaped body, which is bluntly rounded anteriorly, broad at the thorax, and narrows to a pointed posterior. They are most active during the hottest part of the day. Despite being fairly common, they are not often seen as they sit quietly on a branch or twig, flying off with a buzzing sound only when directly approached.

The dorsoventrally flattened larvae have no legs but the thorax is expanded in a characteristic manner. They bore in wood or feed on roots; a few species are leaf-miners, tunnelling between the upper and lower layers of a leaf.

This large family is well represented in Africa. Many species occur in the KNP and are most common where large expanses of wooded vegetation exist.

PLATE 18

Family Dermestidae (Skin beetles, hide beetles)
Velkewers

Dermestids vary in length from a few millimetres to just over a centimetre, and most adults are patterned in rather dull shades of grey or brown.

The only species likely to be seen in the KNP is *Dermestes maculatus* which, when seen from above, is dull brown in colour with white markings much like epaulettes at the base of the forewings. The adults measure little more than 10 mm in length. They are most abundant at carcasses, after blowfly larvae have left, and feed on pieces of drying flesh, sinew or skin. The adults deposit large numbers of eggs at the carcass, and soon thousands of dermestid larvae can be seen gnawing at the skin. If left undisturbed they will consume an entire hide within a matter of weeks. The larvae are dark brown in colour and have tufts of hair covering the body surface. They prefer to remain below the carcass, where they feed at the skin from the inside, thus avoiding the hot rays of the sun.

Dermestes maculatus, a very common beetle at decomposing carcasses, occurs throughout Africa.

PLATE 18

Family Coccinellidae (Ladybird beetles)
Skilpadjies

These dainty, multi-coloured beetles have a characteristic shape, being humped or rounded above and flat below. Their short legs generally do not protrude beyond the side of their body, so that in motion on the stem of a plant the insect resembles a split pea gliding along. Most ladybirds are less than 10 mm in length, and often have the elytra adorned with bright spots or stripes. Most are predatory, both as larvae and as adults, and can be found sitting on plants where they feed on homopterans such as scale insects and aphids.

Common throughout the wooded areas of Africa, especially in areas of lush vegetation, this family is well represented in most regions of the KNP, although they are less common in the drier open grassveld zones.

PLATE 19

Families Lyctidae and Bostrychidae (Powderpost beetles and shot-hole borers)
Houtpoeierkewers en rondekophoutboorders

These two families are discussed together because of their similarity in appearance and habits. The adults are elongate, generally less than 10 mm in length, and dark reddish-brown to black in colour. Shot-hole borers have the head recessed or tucked in below the enlarged pronotum, whereas the head of powderpost beetles can easily be seen from above. Shot-hole borer adults also often have the hind ends of the elytra curving sharply downwards so that a concave depression forms at the posterior end of the body. Along the edges of this depression the elytra often have spinous processes projecting backwards.

The females of both families sometimes lay their eggs in cracks or holes in unhealthy trees, but more often lay them in drying wood, such as trees pushed over by elephants, wind or man. Larvae emerging from the eggs burrow in the sapwood, leaving meandering tunnels densely packed with a powdery dust. When fully grown each larva chews a small chamber for itself, in which it pupates. When the adults emerge they burrow to the exterior to find mates.

The Lyctidae and Bostrychidae occur throughout Africa wherever stands of wooded vegetation are found. They are widespread in the KNP, often invading logs or fallen branches of acacias, marula, kiaat and combretum.

PLATE 19

Family Meloidae (Blister beetles)
Blaartrektorre

Most of the more noticeable species of blister beetle are attractively and brightly coloured to warn predators of their unpleasant taste. They contain in their bodies a substance known as cantharidin, which is distasteful and irritating, causing blistering of the skin. The adults measure between 20 mm and 30 mm in length, are barrel-shaped, and have well-developed legs and elytra with a waxy appearance.

Adults are phytophagous and often can be seen feeding on flowers. Females deposit numerous eggs either in or on the ground, from which emerge tiny, elongate larvae known as triungulins. These are first-stage larvae, and may be recognized by their well-developed legs. The larvae are unusual in that many are parasitic on locust or grasshopper eggs in soil, while others parasitize the larvae and food stores of solitary bees. The triungulins either actively search for the nest or egg-batch of a suitable host, or climb on to a feeding or resting bee and drop off when the host insect reaches its nest. After feeding for a short while, the larvae moult and in doing so change shape, losing the legs and becoming more grub-like and sluggish. After a few more moults the larvae enter the pupal stage, which may be preceded by a non-feeding, resting stage if temperatures are very low. The adults finally emerge to resume the cycle.

Blister beetles are widespread in many parts of Africa. They are locally common in most well-wooded areas of the KNP.

PLATE 19

Family Tenebrionidae (Darkling beetles)
Skemerkewers

A diverse and common family, tenebrionids range in shape from almost round to elongate and from bulbous to flat, and in length from a few millimetres to several centimetres. Most species are dark in colour, generally grey-black, and have the body protected by a well-sclerotized cuticle, and many of the larger species have longitudinal grooves or ridges along the elytra.

Most species are phytophagous both as adults and as larvae, feeding on drying and rotting vegetation, although some will also scavenge opportunistically on dead insects or small animals. Adults are often seen walking on the ground, mostly as solitary individuals, whereas the larvae tend to live in soil or under leaf debris, some species utilizing dung or fungi, or living under bark.

The beetles which are often referred to as 'tok-tokkies' belong to this family. Their name is derived from the sound which results from their habit of tapping the ground with the hind end of the abdomen. Both sexes display this behaviour, which has the function of attracting the opposite sex.

This is a very large family, with representatives occurring in the deserts, savannas and forests of Africa. Many of its species are common throughout the KNP.

Family Scarabaeidae (Dung beetles, rhinoceros beetles and their relatives)

The name Scarabaeidae for most people is associated only with dung beetles; however, the family is very large and includes several subfamilies of which only a few are associated with dung. In size, structure and coloration the beetles vary considerably, but all have clubbed antennae, the last few segments of each antenna being elongated and closely pressed together to form a knob. The larvae are all thick, squat, whitish grubs with a well-sclerotized head having a brown to black colour. They are sluggish in their movements and when uncovered often lie with their bodies bent in a characteristic C-shape.

PLATE 19

Subfamily Scarabaeinae (Coprinae) (Dung beetles)
Miskruiers

These are the true dung beetles. Most of the more abundant species (eg. *Onthophagus*) are only a few millimetres in length and brown in colour. Larger species do occur, however, and these range from the black or brown, medium-sized species which often have rhinoceros-like horns curving up from the head (eg. *Copris* and *Scarabaeus*) and the bright, metallic-coloured species in shades of green and red (eg. *Onthophagus* and *Gymnopleurus*) to the giant-sized species, sometimes reaching 50 mm in length, which are usually black in colour (eg. *Heliocopris*). The larvae and pupae of the Scarabaeinae are favourite items of food for such animals as honey badgers, civet cats and mongooses.

Several genera (eg. *Kheper* and *Scarabaeus*) have dung-rolling habits. Occasionally one or two dung beetles may be seen rolling a ball of tightly compacted dung, one beetle generally doing the rolling while the other assists, walks alongside or rides on the ball. The beetle makes the ball by using its forelegs and the front of its head as scrapers to collect suitable bits of dung at a dung pad, and to compress and shape these. Finally the ball is rolled away, the beetle standing on its forelegs, head down, while making use of its hind legs to push the sphere along. It buries the dung in soil, where the female inspects it, patting and kneading it until it is perfectly smooth and round. She then forms a small hollow in the ball, lays an egg, and covers it with dung in such a way that the originally round ball now resembles a pear, with the egg lying in the smaller, jutting-out end. Having left the burial chamber, the female blocks the entrance with soil and departs. When the larva hatches it eats away at the dung for a number of weeks until it is fully grown, and then pupates within the now hollow sphere. After moulting into the adult stage the beetle breaks its way out of the chamber and moves up to the surface.

A few species make their nests within the dung pad itself. However, most dung beetles tunnel into the soil below the dung pad and excavate a chamber, which they stock with selected bits of dung shaped into small compartments, with each compartment containing an egg. Several species of *Onthophagus*, and the large reddish-brown *Anachalcos convexus*, also feed on the flesh of carcasses in the initial stages of decay.

As is the case with most insects, the value and importance of dung beetles are not always realized; only when they die off or an imbalance is produced in their population levels do we notice their impact on the environment. A very good example of this exists in Australia, where cattle were introduced during the 18th century to feed the growing human population. The climate and vegetation in most parts of Australia proved ideal for cattle and they multiplied rapidly. However, because there were no dung beetles adapted to the texture and consistency of cattle dung in that country, the dung pads remained where they had dropped. Two serious problems then arose: firstly, pasture space was reduced by the millions of dung pads which were being deposited daily and, secondly, disease-transmitting and bothersome flies bred in large numbers in the dung. Entomologists were sent around the world during the 1960s to search for suitable dung beetles to alleviate the problem. A research unit was also established near Pretoria, and many collections of beetles made in the KNP. Several species have now successfully been introduced into Australia and are spreading throughout the areas occupied by cattle.

Dung beetles are abundant and widespread throughout the warmer areas of Africa. They are very common in the KNP, especially at elephant dung where several hundred individuals representing many different species often occur together.

PLATE 20

Subfamily Dynastinae (Rhinoceros beetles)
Renosterkewers

These are fairly large, oblong beetles, the most common species being reddish-brown in colour and having smooth elytra without any sculpturing or ridges. The males have a large 'horn' projecting upwards from near the front of the head, hence their common name. Females have a slight bump in the corresponding position, so that the sexes are easily distinguished.

The larvae – large white grubs – are sometimes seen in compost heaps or piles of rotting vegetation, or feeding underground on the roots of living plants.

Although not often seen due to their nocturnal habits, this subfamily occurs throughout the warmer, well-vegetated areas of Africa. In the KNP adults are occasionally attracted to lights at night.

PLATE 20

Subfamily Cetoniinae

Ranging in size from small to rather large, most species in this subfamily are conspicuously and attractively coloured. Some have white spots on a black background, others yellow wavy lines on a black background, and still others are mottled in shades of red and brown.

The adults, which tend to be diurnal, feed on flowers, fruit and exuding juices on trees. The larvae live in plant debris or decaying vegetation and are therefore not often seen.

A large subfamily and most common in the tropics, the Cetoniinae are widespread in well-wooded areas throughout Africa. Several species occur in the KNP.

PLATE 20

Subfamily Rutelinae

These tend to be medium-sized insects, sometimes brightly coloured but generally dull brown or grey. The adults feed on leaves and fruits of plants, while the larvae prefer the roots of living plants or decaying vegetable material.

The Rutelinae form a rather large subfamily with a wide distribution throughout Africa. Several species occur in the KNP.

PLATE 20

Family Trogidae

A small family of slow-moving beetles, most species are medium-sized and grey in colour. The body is oval in shape, with the underside of the abdomen flat and the head partially deflexed under the thorax. Adults have a strongly sclerotized cuticle, the upper surface of which is deeply pitted and sculptured to give a rough appearance. They are very common at decomposing carcasses, where they feed on the skin, fatty layers below the skin, and occasionally on the blowfly maggots that occur in and around the carcass.

Trogids are found throughout Africa. Several species are common throughout the KNP, hundreds occasionally being found below a single carcass. The most abundant in the KNP are *Trox squalidus*, *T. tuberosus*, *T. melancholicus*, *T. radula*, *T. mutabilis* and *T. rusticus*.

PLATE 21

Family Cerambycidae (Long-horn beetles)
Boktorre, langhoringkewers

Long-horn beetles form one of the largest families of insects, with many highly attractive species. They are easily recognized by their elongate bodies with a pair of long antennae curving away in a crescent, the antennae being at least two-thirds the length of the body. The size of the insect varies considerably, some being small while others may be longer than 60 mm. Although many species are dully or cryptically coloured, many are bright, either uniformly coloured or with lines and wavy patterns adorning the elytra.

The larvae of most species tunnel inside the trunks of dead or dying trees, while some feed on roots and a few invade the stems of living trees. Common especially in the Pafuri area, the large, brown *Macrotoma natala* has large white grubs which tunnel inside decaying *Acacia robusta* and *A. tortilis* trees, a single tree sometimes becoming so infested with these larvae that the inside is riddled with wide tunnels. As the grubs eat their way through the wood, they pass cream-coloured faecal pellets which become tightly compressed in the tunnel behind them. These larvae, known as mabungu grubs, are a great delicacy in some regions when fried, having a taste reminiscent of roasted peanuts.

Long-horn beetles are common throughout Africa wherever woody vegetation exists. Numerous species occur in the KNP.

PLATE 22

Family Chrysomelidae (Leaf beetles)
Blaarvreetkewers

Most of the more commonly seen species in this family are brightly coloured, often with a metallic sheen, and are oblong or oval in shape.

In the larval and adult stages, most live on plant material and are commonly found sitting on shrubs or trees. Some species have aquatic larvae which feed on the roots or stems of water plants, while others have larvae which live within the roots or are leaf-miners of plants on land. The larvae of many species have the habit of attaching the old skin to the hind end of the body after each moult, so that eventually a 'tail' of several shrivelled skins is formed.

The tortoise beetles of the subfamily Cassidinae are somewhat unusual, being round in outline but flattened and having a brilliant, reflective gold colour. These beetles are about 10 mm in length and are occasionally seen on the stems or leaves of shrubs or young trees.

The Galerucinae is a large subfamily, with member species varying considerably in shape and size. During summer of some years a certain dull grey species of *Galerucella* becomes exceptionally abundant and at such times the leaf-feeding larvae may totally defoliate pigeonwood trees (*Trema orientalis*) which, in the KNP, are the favoured food plant of this insect. Many thousands of these yellowish, black-spotted grubs are visible at these times; when fully fed they often form dense clusters at the base of the trunk, where pupae and adults may also be found.

With well over 20 000 species described worldwide, this large family is well represented in most wooded areas of Africa. Many species occur throughout the KNP.

PLATE 22

Family Bruchidae (Seed weevils)
Boontjiekewers

Most of these are small beetles, less than 4 mm in length, and roughly oval in shape. The adults are covered with a dense layer of short hairs which provides a dull, mottled appearance, very effective for camouflage purposes.

The larvae of most species feed on the seeds of pod-bearing plants such as acacia trees, peas and beans. Females lay their eggs on young seed pods, and when the larvae emerge they tunnel through to the seeds inside. Here, each larva feeds on the nutritious endosperm of a seed, finally pupating within the seed. Adults chew their way out.

Although common throughout the KNP, these beetles are seldom noticed because of their small size. A few species are economic pests on crops of the Leguminosae.

PLATE 22

Family Scolytidae (Bark or engraver beetles)
Baskewers

Small beetles (generally less than 10 mm long), elongate and dark in colour, these insects are sometimes considered a subfamily of the Curculionidae (page 107). Although the insects themselves are rarely seen, evidence of their presence is fairly abundant. Most species make tunnels in the bark or in the sapwood of recently dead or dying trees. The most commonly seen 'engravings' are those formed where larvae have emerged from eggs deposited in a central tunnel, and then made their own galleries radiating outwards in a fan shape.

Most larvae feed on the wood while tunnelling; some species allow a fungus to grow in the galleries and then feed on the fungus. Large numbers of bark beetles may attack a particular tree at the same time. This phenomenon is now known to result from the females, or both sexes, giving off a pheromone which causes an aggregation of adults at the tree.

Species of *Ips*, *Dendroctonus* and others are serious pests of forest trees in many parts of the world.

Bark beetles are common throughout the wooded areas of Africa. The 'engravings' of these beetles may be seen on fallen logs and standing dead trees throughout the KNP.

PLATE 22

Family Curculionidae (Snout weevils)
Snuitkewers

This is the largest family in the animal kingdom. The adults vary considerably in shape, but most are oblong. In size they range from small to fairly large, and are generally dull grey or brown in colour. They have an unusually well-sclerotized cuticle which makes them extremely hard and tough. The characteristic feature of these insects is the elongated rostrum, which also has a pair of 'elbowed' antennae jutting from the sides. The basal part of the antenna consists of a single long segment, followed beyond the 'elbow' by a number of smaller segments. Although most curculionids have a distinct and easily recognizable rostrum, this is not always obvious in some species.

Weevils feed on plant material, many species being serious economic pests in most countries of the world. Most females deposit their eggs in the tissues of the host plant, chewing a hole into the plant and laying their eggs inside. Female adults tend to have a longer rostrum than males, possibly due to their egg-laying behaviour. The larvae are typically grub-like in shape, are legless, and feed on the plant tissues.

In an effort to combat invasive exotic weeds which have established themselves in the KNP, researchers in the Department of Agriculture and Water Supply have done much work to introduce natural enemies of these plants. These enemies are brought from their countries of origin, where the weeds are kept under control through the action of the insects in a classical example of biological control. Among other insects introduced in this manner are the weevil species *Neohydronomus pulchellus*, which feeds on water cabbage (*Pistia stratiotes*), *Cyrtobagus salviniae*, which feeds on Kariba weed (*Salvinia molesta*), and *Neochetina eichorniae*, which is destructive to water hyacinth (*Eichhornia crassipes*). Two other weevils, *Neodiplogrammus quadrivittatus* and *Trichapion lativenne* have been released immediately outside the KNP, along the banks of the Sand River, in an attempt to control *Sesbania punicea*. It is hoped that these two weevil species will prove themselves to be effective biocontrol agents of the alien *S. punicea* plants.

Snout weevils are widespread and common throughout Africa. They are found in the KNP wherever shrubby vegetation occurs.

PLATE 23 | **Family Brentidae** (Straight-snouted weevils)
Valssnuitkewers

This is a small family of beetles which closely resemble curculionids (page 107) in appearance and habits, but are easily distinguished by the fact that their antennae are not elbowed, as are those of curculionids. Most brentids are narrow and elongate in shape. Most species are plant-feeders and have habits similar to those of the curculionids, but a few are predators on other insects. Females tend to chew holes into plant tissue and deposit their eggs inside.

Straight-snouted weevils are confined largely to the tropics, occurring in most of the warmer areas of Africa where wooded vegetation occurs. They have been collected at several localities throughout the KNP.

PLATE 23 | **ORDER STREPSIPTERA** (Stylopids)
Krenkelvlerke

This is a small order of highly modified, unusual insects, and even entomologists consider themselves fortunate when having the rare opportunity of seeing a specimen. All species are parasitic. Most measure between 1 mm and 4 mm in length. Adult males have distinct, branched antennae, large eyes which bulge from the head, the first pair of wings modified into club-like structures, and large, fan-shaped hind wings with few veins. Adult females are wingless and often have neither legs nor antennae: these appendages are unnecessary as females are endoparasites within a wide range of other insects.

Each female gives birth to a large number of larvae (the eggs hatch inside the female) which have well-developed legs, and these 'triungulin' larvae move out in search of insect hosts. Once having entered a host, the larva moults and in the process loses its legs. The larvae pass through several instars and pupate within the host. Adult males leave the host after emergence from the pupa, but generally live for only a few hours.

Stylopids are widely distributed over Africa and other parts of the world. Very few specimens have been collected in the KNP, but this may be a reflection of their specialized lifestyle and because no specific search has been made for them.

PLATE 23 | **ORDER MECOPTERA** (Scorpionflies or hanging flies)
Skerpioenvlieë

Most mecopterans are slender, medium-sized insects, orangey to light brown in colour, and have elongate wings and abdomen. They have long, thin legs and a head which is extended in front into a short, pointed rostrum. These features cause them to closely resemble crane flies (Diptera: Tipulidae), so that at a quick glance they are difficult to tell apart. Closer examination will reveal the four wings and tibial spurs (spiny projections near the ends of all the legs) which crane flies lack.

In an African context, the name 'scorpionfly' is misleading; it originates from certain northern hemisphere species which have an upcurved abdomen ending in a bulbous tip, much like a scorpion's sting. Only the family Bittacidae occurs in Africa, and none of its members have this abdominal feature. Locally, a more accurate common name would be 'hanging fly' as they have the habit of suspending themselves from twigs using their front and middle pairs of legs. All African hanging flies are predacious and generally feed on other insects.

The mating habits of these insects are rather interesting. Like some other insects and many spiders, the female is very likely to seize and eat the male if she has the opportunity. To escape this fate, hanging fly males capture an insect, which is then presented to the female to occupy her while the male proceeds with copulation. The larvae have secretive habits, as is the case with the immature stages of most insects, and are rarely seen. They resemble the caterpillars of moths and butterflies, but hanging fly larvae can be distinguished by their eight pairs of fleshy legs below the abdomen, and 20 or more simple eyes clustered on either side of the head. The larvae are usually found crawling in shaded vegetation close to the ground, and when fully grown dig into the soil to pupate.

More than 50 species have been recorded, from forested areas to open grassveld, in most parts of Africa. They are likely to occur in many regions of the KNP, but have been collected only in the southern half.

ORDER DIPTERA (Flies)
Vlieë

The Diptera ('two-winged' in Latin) form a very large order of highly specialized insects. The two-winged feature distinguishes flies from most other insects: the hind wings are modified as halteres, these being visible as small, club-like structures projecting from the thorax just behind the forewings. Many species have a considerable impact on human life. In terms of medical and veterinary importance, flies surpass all other orders of insects – indeed, all other animals – as carriers of disease to man and animals, causing death, suffering and financial loss to many millions of people every year. Malaria, yellow fever, leishmaniasis, sleeping sickness, elephantiasis, onchocerciasis and anthrax are only some of the maladies known to be transmitted by flies. But many species are also of great benefit to us, as pollinators of plants and especially as predators and parasites on some of our most important agricultural crop pests.

Flies vary considerably in shape and size. They also exhibit a great range in habits, many being associated with water and many with dry land, some being scavengers, others feeding on blood and some catching insects. The larvae, which are maggot-like in shape, tend to live in concealed and secretive situations, as do the immature stages of most insects.

SUBORDER NEMATOCERA

These are elongate, slender flies, easily recognizable by their 'pearls-on-a-string'-like antennae. Their legs are generally long and thin. These flies are the most primitive members of the Diptera.

PLATE 24

Family Tipulidae (Crane flies)
Langpote

Crane flies can most easily be distinguished by their exceptionally long, fragile legs. They are occasionally mistaken for mosquitos (page 110), which they superficially resemble. They lack the long, thin, piercing mouthparts of mosquitos, however.

The larvae are elongate and slender, and make use of a wide variety of habitats. They can be found in rotting vegetation, under leaf-litter, in soil feeding on the roots of grasses, and above soil among mosses and rocks. Some species have aquatic larvae, some of these feeding on algae while others prey on small, water-dwelling animals.

Crane flies are fairly common in all the warmer, well-vegetated areas of Africa. They have been recorded throughout the KNP.

PLATE 24

Family Psychodidae (Moth-flies and sand-flies)
Motvliegies

Rarely noticed because of their small size (generally less than 5 mm long), these flies nevertheless are fairly common in places such as public toilets, drains and sewage works. The numerous fine hairs covering the broad wings cause them to resemble small moths. They fly weakly, for short distances. The larvae live in water and feed on algae, fungi and bacteria, and are often an important factor in the purification process at sewage works.

The best-known members of this family are those species responsible for the transmission of various forms of leishmaniasis. Caused by intracellular protozoan parasites, this widespread disease manifests itself in several ways: cutaneous leishmaniasis, resulting in boils and lesions on the skin, is referred to as 'oriental sore'; visceral leishmaniasis is generally referred to as 'kala-azar'. Reservoirs of this disease include certain rodents, rock hyraxes (dassies) and dogs; humans become infected if they are bitten by flies which have fed on these sources. Leishmaniasis is not known to occur in the KNP.

Members of the subfamily Phlebotominae tend to be very small, are silvery in appearance and hold their wings above the body. The larvae live in moist soil where they feed on rotting vegetation. Only the females in this subfamily are blood-suckers, and it is only those members of the genus *Phlebotomus* which feed on mammals that are potential vectors of leishmaniasis. Species of the genus *Sergentomyia*, in the same subfamily, are ectoparasites on animals such as frogs, snakes and lizards, and sometimes act as vectors of trypanosomes. *Sergentomyia squamipleuris* has been found parasitizing the frog *Bufo pusillus* in the Pafuri area of the KNP.

Several genera and species of sand-flies have been collected in the Pafuri area, but they appear to be rarer farther south in the KNP.

PLATE 24

Family Culicidae (Mosquitos)
Muskiete

Adult mosquitos have moth-like scales scattered on the body and along the veins of the wings. Their mouthparts are modified into an elongated, piercing proboscis. As in the Chironomidae (page 112), the mosquito males have plumose or feathery antennae, while those of the female are narrow with short bristles.

Only female mosquitos are blood-sucking, the protein obtained in this way being necessary for maturation of the eggs in many species. The males – and in some species, both sexes – feed on nectar or juices obtained from rotting fruit. Most mosquitos are nocturnal, this being an adaptation corresponding to the period when humans and many other vertebrates are least active and therefore most susceptible to attack.

The females of most species lay their eggs in water. Some species lay drought-resistant eggs in holes, and these hatch when the holes are filled by rain. Females of *Anopheles* and *Aedes* lay their eggs singly, but other mosquitos bind their eggs together with a secretion and these float raft-like on water. The larvae in most cases are elongate, dull brown in colour, and have a respiratory siphon or breathing tube at the hind end of the body. They suspend themselves by hairs at the surface of quiet pools of water, occasionally wriggling down to the bottom, then swimming back to the surface with vigorous movements. Pupae spend their time in pools in a similar manner. When the adults are ready to emerge, a split develops along the upper side of the pupa, through which the adult escapes.

Mosquitos are well-known as important vectors of a wide range of diseases, including malaria, yellow fever, elephantiasis and other filarial diseases, and many arboviruses. Malaria ranks as the single disease which disables the most people throughout the world each year, resulting in more deaths than most other diseases; in Africa alone, more than a million children die each year from malaria. The disease is caused by microscopic sporozoans of the genus *Plasmodium* which are taken up with the blood when mosquitos feed on infected humans. These parasitic organisms then move through the gut to the salivary glands of the mosquito, and are injected into the next human when the mosquito feeds again. Only some members of the subfamily Anophelinae are effective vectors of malaria. *Anopheles* mosquitos are easily recognized by the upright angle of their body when at rest, the head being depressed and the abdomen pointed away from the substrate at a roughly 45-degree angle. (Most other mosquitos hold their bodies parallel to the substrate.) The indoor biting species *Anopheles gambiae* and *Anopheles funestus* were the main transmitters of malaria in Africa in previous years, but these species have been virtually eradicated by a continuous control campaign by state health officials. The outdoor biting species *Anopheles arabiensis* is considered to be the main vector at present. Malaria is still prevalent in many parts of southern Africa. Cases regularly occur in the KNP and adjoining Lowveld areas, especially during the wet summer months. Visitors are strongly urged to take anti-malaria tablets when visiting the area during all but the late winter months.

Another disease, although fairly rare, transmitted by mosquitos in the Lowveld is the debilitating fever called chikungunya. It is known that monkeys and baboons serve as reservoirs for this virus, and that it is occasionally transmitted to humans by *Aedes furcifer*. Symptoms of the disease are fever, headache, muscular and joint pains, skin rash and swollen lymph glands.

Particularly fond of biting during the day, the black-and-white striped *Aedes aegypti* is responsible for transmitting yellow fever in many parts of the world, but fortunately this disease has not established itself in South Africa.

Mosquitos are common throughout the KNP, especially during summer. They are most abundant alongside rivers, streams, waterholes and pans, although they do occur elsewhere due to the habit of some species of using as breeding sites temporary pools or rainwater collected in concave rocks, gutters and hoof prints.

PLATE 24

Family Ceratopogonidae (Biting midges)
Bytende muggies

Small flies, often only 1 mm in length, the members of this fairly large and widespread family have a diverse range of habits in both the adult and larval stages. The females of most species are blood-sucking ectoparasites on vertebrates and insects. The males – and females of many species – live on plant juices and are often vitally important as pollinators. The larvae are usually found in aquatic environments such as in rivers, pools or marshes, or in fairly moist situations such as in shaded, decaying vegetation.

Most of the blood-sucking species which attack vertebrates occur in the genus *Culicoides*, and many are frequent parasites on humans. In Africa south of the Sahara between 20 and 30 viruses have been isolated from *Culicoides* midges. Two species, *C. austeni* and *C. grahamii*, are known to transmit the filarial worms *Dipetalonema perstans* and *D. streptocera* to humans in tropical Africa. In South Africa, *Culicoides imicola* and others are of great importance as carriers of bluetongue virus to sheep, and the virus causing African horse-sickness.

Biting midges are common and widespread in most areas of Africa. During one week of collecting in the KNP, De Meillon and Wirth (1981) reported having collected 47 species from 13 genera, 15 of these species being new to science; they write that the 'ceratopogonid fauna of the KNP is obviously a very rich one...'. This was confirmed by the collections of Rudi Meiswinkel of the Onderstepoort Veterinary Research Institute and the author during the mid-1980s: a total of 87 species was recorded at Skukuza, of which 38 were of the genus *Culicoides*. This figure for *Culicoides* is the highest number of species known from a single locality in South Africa. Skukuza also yielded 10 of 18 species of the subgenus *Avaritia*, also the highest diversity of this vector group per sampling site in South Africa. Diurnal biting midges of the genera *Forcipomyia* (subgenus *Lasiohelea*) and *Leptoconops* have also been recorded from the KNP.

PLATE 25

Family Chironomidae (Midges)
Muskietmuggies

Midges are very similar to mosquitos (page 110) in both size and appearance, but can easily be distinguished by the lack of the long, piercing proboscis and the fact that they do not have moth-like scales on the wing veins. Adults are generally uniformly coloured in pale shades of green, brown or yellow. They are associated with moist situations, occurring most commonly along rivers, and in streams, large pools and marshes.

The larvae live in water, where they feed on organic debris, and they form an important component in the diet of fish and other aquatic organisms. Some species are exceptional in that they have haemoglobin in their blood for gas transport, just as humans and other vertebrates do. The pupal stage is also spent submerged; when ready to emerge as an adult, the pupa wriggles to the surface where the adult exits through a split along the upper surface of the pupal skin.

Chironomids are common throughout Africa. In the KNP they are often attracted to lights on summer evenings.

PLATE 25 | **Family Cecidomyiidae** (Gall gnats or gall midges)
Galmuggies

Cecidomyids are generally dull in colour and measure a few millimetres in length. The body and wings are covered with a fairly dense coat of fine hairs, and the wings have a characteristic reduced venation.

The larvae of certain species are responsible for causing growths on the leaves or stems of plants, which has given rise to the common name: the gall is a growth response of the plant to chemical substances injected into t by the larva, and results in a protective covering within which the larva lives. When fully grown the larva pupates within the gall, the adult eventually emerging via a narrow tunnel at the base of the gall. The large, spiky galls formed on silverleaf terminalias (*Terminalia sericea*) are caused by cecidomyid midges; if a gall is sliced open, numerous maggots will be seen inside. These midges have been identified as *Tetrasphondylia terminaliae*. However, not all plant-feeding species create galls. Some species of gall midge larvae live in rotting vegetation, while others are parasites or predators on insects.

Gall midges are common in most areas of Africa. They are found throughout the KNP, where they are often attracted to lights on warm summer evenings.

SUBORDER BRACHYCERA

Flies in this suborder tend to be more compact in structure than the Nematocera (page 109), and have shorter, stouter legs. The antennae are much shorter, often almost invisible to the human eye, and generally have fewer than eight segments. The antennae are often of the aristate form, meaning that they have two stout basal segments, followed by an elongate segment to which is joined a very thin terminal segment.

PLATE 25 | **Family Tabanidae** (Horse flies)
Blindevlieë

Medium to large in size, these flies are generally patterned in shades of grey and brown, although some species display combinations of black and white. The body often tapers to a pointed posterior, and the eyes in both sexes are large, those of the males being holoptic.

Adult males generally feed on nectar, while the females suck blood from mammals, using their sharp mouthparts to rasp or cut through the skin. Because the females require several blood-meals during their life span and obtain these from different hosts, they are important transmitters of disease in animals and man. Several members of the genus *Tabanus* are known to carry surra disease (*Trypanosoma evansi*) in horses, and *Chrysops dimidiata* transmits the parasitic eye worm *Loa loa* in humans in West Africa. Tabanids are also known to transmit anthrax in animals and nagana (*Trypanosoma brucei*) in cattle.

The larvae are carnivorous and most live in damp situations such as in moist soil or rotting vegetation. Some species are aquatic, preying on small animals sharing their habitat.

Generally, tabanids are common throughout the warmer areas of Africa, excluding the more arid desert areas. Stuckenberg (1974) conducted surveys in the KNP on the distribution and abundance of horse flies, and compiled a list describing 30 species from 11 genera.

The genera are *Chrysops*, *Thriambeutes*, *Sphecodemyia*, *Jashinea*, *Tabanocella*, *Philoliche*, *Ancala*, *Atylotus*, *Tabanus*, *Mesomyia* and *Haematopota*. Many of the species are widespread and abundant.

PLATE 26

Family Asilidae (Robber flies)
Roofvlieë

With a few exceptions, robber flies are elongate in shape, and have a fairly long, slender abdomen and well-developed legs. The wings at rest are held flat over the abdomen. They have a characteristic depression between the eyes on top of the head. Projecting downwards from the head are the well-developed, triangular mouthparts. Characteristic of robber flies is their 'bushy-beard' appearance, given by the hairs which so abundantly adorn the lower and front side of the head. These insects vary from 10 mm to over 30 mm in length, and are generally dull brown to grey in colour. Several species are handsomely patterned, however, such as the brilliant coppery *Hoplistomerus*, recognizable by the wavy black band on the wing, while other species are boldly coloured in black and white.

All adults are predatory, generally taking other insects – often much larger than themselves – on the wing. Diurnal, they are most frequently seen singly, sitting quietly on a twig, waiting patiently for a suitable prey insect to fly past. The larvae live in soil and rotting vegetation, where they prey or scavenge on insects and small animals, but a few species feed on plant material.

Species of *Hyperechia* mimic in shape and colour the large carpenter bees of the genus *Xylocopa* (page 141), and also prey on these formidable bees. Larvae of *Hyperechia* apparently also live in the same burrows as *Xylocopa* larvae, but it is unsure if the larvae prey on those of *Xylocopa*.

Robber flies are commonly seen throughout Africa wherever other insects occur.

PLATE 26

Family Bombyliidae (Bee flies)
Byvlieë

Based on the number of species currently described, the Bombyliidae is the largest family of Diptera in Africa. Bee flies are generally squat, medium-sized insects, often having the body covered in a dense layer of short hairs. When at rest the large forewings are held horizontally on either side of the body. In most species the mouthparts are modified into a long, thin proboscis which projects from the front of the head. Adults have a rather characteristic flight: a short, rapid burst followed by a period of hovering.

Adults are most commonly seen on warm, sunny days, feeding on nectar at flowers – they appear to prefer those near ground level – or resting on the ground. They are most common at the edges of riverine forests or in wooded areas where flowers occur.

The larvae are parasites on the immature stages of a large range of insects, including grasshoppers, moths, wasps and bees. Adult females drop their eggs singly at suitable sites, and when these eggs hatch the larvae crawl off into the hosts' nest, consuming any stored food first (in the case of wasps or bees) before preying directly on the larvae or pupae.

Bee flies occur in suitable habitats throughout Africa. They are common in the KNP.

SUBORDER CYCLORRHAPHA

When fully grown the larvae of the suborder Cyclorrhapha pupate within a puparium which is formed from the skin of the last larval stage which has become hardened and shrunken. The antennae of the adults are usually three-segmented (or appear three-segmented), with an arista projecting from a dorsal position.

PLATE 26

Family Syrphidae (Hover flies)
Sweefvlieë

The most characteristic feature of hover flies, the 'spurious vein', unfortunately is meaningless to most people except entomologists. In general appearance, however, hover flies are smallish to medium-sized insects with an elongate body, and most species are brightly coloured either with spots or banded patterns, yellow and black being the predominant colours.

Adults feed on nectar. Like bee flies (page 114), they can often be seen hovering at or near flowers, but they also have the habit of hovering in small clearings or above underlying shrubbery. Many species mimic bees and wasps in coloration and sound. The larvae occupy a range of habitat types, some being scavengers – a few in the nests of ants – while others are predatory. The larvae are slug-like or maggot-like in appearance and can occasionally be seen feeding on aphids on twigs or leaves.

Hover flies are common in well-wooded areas, along river banks and near streams or pans throughout Africa. Many species occur commonly in the KNP.

PLATE 27

Family Lauxaniidae

Most lauxaniids resemble small house flies (see below) in body shape and are patterned with spots or bands. The adults are generally found in wooded areas, especially along river banks or in other moist situations. The majority of species have larvae which live in decaying vegetation, while some occur in birds' nests.

The family is found throughout Africa. In the KNP they are common in suitable habitat.

PLATE 27

Family Muscidae (House flies and their relatives)
Huisvlieë

Although the house fly, *Musca domestica*, is perhaps the best known member of this large family, the Muscidae contains many other species of medical and veterinary importance. While they range from very small to fairly large in size, many are very similar in appearance to the house fly.

Most muscids have a soft, retractable proboscis which they use to feed on a wide range of liquid substances. Members of the subfamily Stomoxyinae and several of the Muscinae have long, piercing mouthparts, or small teeth at the end of a hardened proboscis, which are used to pierce or scratch the skin of vertebrate hosts to obtain the blood on which the flies feed. The majority of the larvae live in dung, carrion or other decaying matter, or prey on other fly larvae.

The life cycle of the house fly is representative of many species in this family. The female house fly deposits her whitish, elongate eggs in batches of 100-150 in dung or other organic refuse. The resulting larvae generally take from five to eight days (in summer) to complete their development. If the sun has dried the dung sufficiently the larvae may pupate inside it, otherwise they will find a concealed position nearby, such as under leaves, stones or similar shady objects. The skin of the last larval stage becomes shrivelled and hardened into a smooth, reddish-brown, barrel-shaped puparium within which the pupa forms. After four to seven days the pale, soft adult emerges from the puparium, using a haemolymph-filled sac in front of the head to push open the 'cap' of the puparium. The fly then rests until the body and wings have hardened, and flies off. In winter the process slows down, taking roughly twice as long.

The house fly subspecies *Musca domestica calleva* is widespread throughout the KNP. Many other species of *Musca* are common in the KNP, including *M. tempestatum*, which abounds on and around blue wildebeest and breeds prolifically in the dung of these animals.

Musca sorbens is also common, and elsewhere in Africa this species, which has the habit of lapping up moisture around the eyes of animals and man, has been found to transmit trachoma and other ophthalmias.

Musca domestica and its close relatives are known to be carriers of numerous diseases for example anthrax, poliomyelitis, cholera, bacillary dysentery, typhoid fever, salmonellosis, leprosy, yaws, trachoma, infective hepatitis and foot-and-mouth virus, as well as various leishmanias, trypanosomes, trichomonads, platyhelminths, nematodes and mites. The small or lesser house fly, *Fannia canicularis*, also occurs in the KNP, but is less noticeable and less common than the true house fly.

Haematobia spinigera and *H. thirouxi potans* are small, blood-sucking flies which occur on several animals but are especially abundant on buffalo, with several hundred to more than a thousand flies commonly found on a single buffalo. Large numbers of these flies breed in buffalo dung. Species of *Haematobosca* also commonly suck blood from buffalo and other animals, as do several species of *Stomoxys*, including the well-known stable fly, *Stomoxys calcitrans*.

The blood-sucking *Stygeromyia maculosa* has been recorded in low numbers in the KNP. *Rhinomusca dutoiti* is a common blood-sucking fly on both white and black rhino, and appears to be confined to these hosts. As both these animals were extinct in the area for many years, these flies must have accompanied their hosts, probably as eggs in dung in the transporting crates, when rhino were reintroduced into the KNP in the 1960s and '70s.

The larvae of the genus *Passeromyia* are parasitic on bird nestlings. The female fly deposits her eggs in a bird's nest, and upon hatching the larvae crawl on to the nestlings to feed on their blood, occasionally killing the nestlings. When fully grown the larvae pupate inside the nest. *Passeromyia heterochaeta* is widely distributed in Africa and has been recorded from the nests of many bird species. It has not been recorded in the KNP, but Zumpt (1966) does mention it occurring in the Transvaal (localities uncited).

PLATE 27 | **Family Gasterophilidae**

This is a fairly small family of flies, the larval stages of which are endoparasitic in various mammals. The adults have non-functioning, rudimentary mouthparts, and most are rather squat and, often, hairy. They tend to live for only a few days and are not often seen. Four subfamilies are recognized.

The larvae of the Gasterophilinae are parasites in the alimentary tract of equids and rhinoceros, and two genera have been described. Females of the genus *Gasterophilus* lay their eggs on the skin of the host and the first-instar larvae are either licked up by the host or find their own way into the mouth of the host; once inside, they burrow into the mucosal layers. The second- and third-stage larvae attach themselves by their mouth-hooks to the lining of the pharynx, stomach or hind gut of the host, remaining there for several months and feeding on exudations or blood.

When fully grown the larvae are excreted by the host with the faeces, following which they dig into the soil to pupate. Several species have been found parasitizing Burchells' zebra in the KNP, these being *Gasterophilus haemorrhoidalis*, *G. inermis*, *G. meridionalis*, *G. nasalis*, *G. pecorum* and *G. ternicinctus*.

The adults of the other genus, *Gyrostigma*, are very large, often reaching 25 mm in length. The larvae are parasitic in the stomachs of rhinoceros and have the same habits as *Gasterophilus*. *Gyrostigma pavesii* is widespread throughout Africa as a parasite in both the black and the white rhinoceros, and is commonly collected from white rhino in the KNP.

Adults in the subfamily Cobboldiinae vary considerably in external appearance. The larvae, however, are all similar in habits and appearance. They are endoparasites in the alimentary tract of African and Indian elephant. The females lay their eggs at the base of the tusks of the host and the larvae complete their development in the stomach of the host. When fully grown they work their way back up to the mouth, and probably drop out while the elephant is feeding (Zumpt, 1970). *Cobboldia loxodontis* is widespread in Africa in the stomach of elephant, and occurs in large numbers in the KNP in the same host. The adult flies measure 10 mm to 14 mm in length and have a metallic-bluish body.

The larvae of another fly, *Cobboldia chrysidiformis*, were also recovered from elephant in the KNP by Zumpt (1970), in very low numbers. This record needs to be verified; numerous attempts by the author to rear adults of this species from larvae randomly collected from elephant throughout the KNP have failed. The only other recorded locality for this species is the Congo.

Only one species is contained in each of the subfamilies Rutteniinae and Neocuterebrinae, both of which parasitize the African elephant. The larvae develop in skin boils or dermal pockets at various sites on the host. Neither species has been found in the KNP, the only recorded localities being Cameroon and the Congo (Zumpt, 1970).

PLATE 27 | **Family Oestridae** (Warble flies or bot flies)

Like the Gasterophilidae (see above), the adults of these flies are stoutly built and often hairy, have non-functioning, rudimentary mouthparts, and are short-lived and rarely seen. The larvae are all parasitic on a range of mammals, living either in the naso-pharyngeal passages or in the skin. Two subfamilies are recognized, based on the habits of the species in these two groups.

Adults of the subfamily Oestrinae give birth to live young which develop in the nasal or pharyngeal cavities and passages of various ungulates. When fully grown the larvae find their way out of the host, generally being sneezed out, to pupate in the soil or among surface litter. The genus *Pharyngobolus* contains a single species, *P. africanus*, whose larvae attach themselves to the upper oesophageal regions of the African elephant. Specimens have been collected in many parts of Africa as far south as Zimbabwe, but have not yet been found in South Africa. The larvae of the genus *Oestrus*, which has several species widely distributed in Africa, feed on exudates in the frontal sinuses and nasal passages of various Bovidae; the well-known sheep nasal bot fly, *Oestrus ovis*, which has an almost cosmopolitan distribution, belongs to this group. The larvae of *Oestrus aureoargentatus* and *O. variolosus* occur commonly in blue wildebeest in the KNP, and also in other mammals such as roan antelope and tsessebe. Only two species occur in the genus *Gedoelstia*. Although the larvae are normally found in the nasal cavities and frontal sinuses of certain antelope of the tribe Alcelaphini, these insects sometimes attack sheep or other domestic animals, in which they cause oculo-vascular myiasis ('*uitpeuloog*'), a disease which arises as a result of their abnormal development in a host species to which they are not well adapted. Both *Gedoelstia cristata* and *G. haessleri* have been found in large numbers parasitizing blue wildebeest in the KNP. Mixed populations of *Oestrus* and *Gedoelstia*, together with other genera such as *Kirkioestrus* and *Rhinoestrus*, often occur within the same host animal.

In contrast with the Oestrinae, females of the subfamily Hypoderminae are oviparous, and it is the second- and third-instar larvae that develop in the skin of various mammals. No member of this subfamily has been recorded from the KNP, although *Strobiloestrus clarkii* has been recorded parasitizing the common reedbuck in Natal and is known from other mammals such as klipspringer, steenbuck, mountain reedbuck and kudu, all of which occur in the KNP, rendering it possible that this species may yet be found here.

PLATE 28

Family Calliphoridae (Blowflies)
Brommers

This is a large and diverse group of insects whose members vary considerably in appearance. Most of the more common species are 10 mm or less in length, and have a squat, compact body. Coloration differs widely, ranging from orange to charcoal black, although many species are metallic-green or -blue. The origin of the term 'blowfly' is obscure. William Shakespeare in at least two of his works refers to meat having been 'blown' by flies, alluding to the habit of some calliphorids of depositing their eggs on fresh meat. It seems likely, therefore, that the common name has its origins in this Old English colloquial term.

The larvae may be found either developing in decomposing organic matter, living endoparasitically within other invertebrates, or feeding as ectoparasites on the blood or healthy tissue of living mammals; the adults are not parasitic.

Two species which are commonly seen at fresh carcasses in the KNP are *Chrysomya marginalis*, a large fly with a brilliant, metallic-blue body and large red eyes, and the somewhat smaller *C. albiceps*, which is metallic-green and has thin, black bands across the body. These two are the first to arrive in the succession of insects which utilize dead animals, the females laying their eggs in the mouth, nostrils or any other shaded crevice on the carcass. The larvae are extremely efficient as scavengers, being able to strip a medium-sized carcass, such as that of an impala, of all its soft tissues within four to six days (in summer). It has been established that these insects play an important part in disseminating the bacterial disease anthrax from dead animals, by defecating droplets of infected refuse on to leaves which other mammals may consume.

In general, however, blowflies are of great benefit as scavengers which remove dead animal matter, especially in riverine forests or other densely wooded areas where carcasses often lie undiscovered by vertebrate scavengers. Another metallic-green fly, closely resembling *C. albiceps* in size, shape and colour, is the fairly common *C. putoria*. The larvae live in faeces, so these flies are most common in areas where open pit-latrines still exist.

Lucilia cuprina, also similar to *Chrysomya albiceps* in size and colour, occurs in the KNP in low numbers. It is best known for its importance in producing myiasis in sheep, resulting in 'sheep-strike', a condition in which the larvae live in moist wool and gnaw at the skin of the sheep, feeding on the serous exudations. If left unchecked the sheep lose condition from the constant irritation, develop sores which attract secondary flies, and eventually die. In a natural system such as that in existence in the KNP, the females of this species deposit their eggs at carcasses, but the larvae are out-competed by those of *Chrysomya*, so that only a few attain adulthood. *Lucilia sericata*, commonly recorded in the KNP, very closely resembles *L. cuprina* in appearance but is not important as an agent of sheep-strike. The larvae are occasionally found in wounds on mammals, but they restrict themselves to feeding on the rotting tissue, and thus they have a beneficial effect: during the First World War surgeons would deliberately introduce sterile larvae of this species into the festering wounds of soldiers, where the maggots rapidly consumed the infected tissue while pouring out digestive secretions which contain powerful anti-bacterial enzymes, leaving a clean wound which healed rapidly.

Found throughout tropical and subtropical Africa, and occasionally at Pafuri in the KNP, the larvae of *Auchmeromyia luteola* (known as Congo floor maggots) parasitize humans where living conditions are primitive and people have the habit of sleeping on the floor. The female fly, orange-brown in colour, lays her eggs in crevices in the floor. The larvae then crawl out at night to gnaw at the skin of sleeping humans, feeding on the blood. Although man is a major host of this species, it has also been found in the burrows of warthogs and antbears. *A. bequaerti* is similar in appearance to *A. luteola*, and its larvae are ectoparasites on warthogs and antbears. The adults are commonly found throughout the KNP, especially in the Pafuri area.

Also known as the putsi fly or tumbu fly, *Cordylobia anthropophaga* is widely distributed in Africa, including South Africa, and frequently causes human myiasis. It is most active during the rainy season, when the females lay their eggs in soil moistened by urine or excreta, or on moist clothing hung up to dry. When the larvae come into contact with humans they burrow under the skin, leaving only the posterior visible. They complete their larval development in this position, finally dropping to the ground where they pupate.

Man is not the most common host of these larvae, a wide range of other mammals usually being parasitized. Domestic dogs are frequently utilized by the maggots, but other hosts include rodents, vervet monkeys and even leopards.

In the KNP occasional outbreaks of human myiasis have occurred, and in the warm summer months visitors are advised to iron all items of clothing, especially babies' nappies, as this will kill any eggs or larvae.

PLATE 28

Family Sarcophagidae (Flesh flies)
Vleisvlieë

These are familiar to most people as elongate, grey flies which occasionally buzz loudly into houses or on to exposed, raw meat. Most are medium-sized and have a characteristic checkerboard-patterned body with an overall grey appearance.

Although most flesh flies breed in dung – commonly elephant dung – many will also breed in meat if given the opportunity. The females give birth to small larvae which burrow into the food source. Covering meat with netting does not help as the larvae work their way with surprising ease through the mesh and on to the meat .

Flesh flies are found throughout Africa. They are common in the KNP.

PLATE 29

Family Tachinidae

These are stout, squat, blowfly-like insects, generally about 10 mm in length and often yellowish-brown in colour. Their characteristic feature is the very bristly appearance of the body, and of the hind abdomen in particular. In the larval stages, tachinids are parasitic on a wide range of other insects and other arthropods such as spiders. Caterpillars in particular are heavily parasitized.

The female adult attaches the eggs by means of sticky secretions to the outside of the host, or she may pierce the body of the host and lay the eggs inside, or deposit the eggs on the food source of the host, with the host then swallowing them with the food. The first-stage larvae are typically parasitic in that they feed on nutrients in the host's body without greatly impairing its normal functions. In the third and final stage the larvae become more predatory in their feeding habits, destroying vital body tissues.

When ready to pupate the larvae eat holes through the body wall of the host and in most cases drop to the ground, where they form pupae. The host usually dies.

Tachinids are fairly common in most well-vegetated areas of Africa. They are widespread in the KNP, where they are most often seen in the undergrowth of wooded areas such as riverine forests.

PLATE 29

Family Glossinidae (Tsetse flies)
Tsetsevlieë

Tsetse flies are generally brown, about 10 mm in length, and have a rigid proboscis projecting forwards from below the head. More than 20 species have been described, all of them restricted to Africa, and all are blood-sucking as adults on various mammals, birds and reptiles.

Females do not lay eggs. A single larva develops inside the reproductive tract of the female, being fed there from secretions supplied by well-developed accessory glands. When fully grown the larva is deposited on the ground, where it digs into the soil to pupate.

Members of the genus *Glossina* are well known as vectors of various species of *Trypanosoma*, some of which cause sleeping sickness in man and nagana in domestic mammals such as cattle, horses, dogs, sheep and goats. Tsetse flies pierce the skin of an infected host with their hardened proboscis, take up the microscopic blood-parasites with their meal and infect the next host when feeding again. If passed on in this way to humans and domestic mammals – which, incidentally, are not the preferred hosts of these flies – the trypanosomes frequently cause a severely disabling or fatal infection. Several species of wild mammals serve as reservoirs for these trypanosomes and, by having adapted and developed tolerance for these parasites, are not adversely affected by their presence.

Although tsetse flies are not found in the KNP today, many historical accounts indicate that the previous distribution of *Glossina morsitans* and *G. pallidipes* did include certain belts passing through this region. The massive rinderpest outbreak of 1896 that killed large numbers of the flies' host animals in the KNP, most importantly buffalo and warthog, was probably the major cause of the tsetse fly population dwindling and eventually dying out in the region. Subsequent control measures by the Mozambican and then-Rhodesian governments caused the population to decrease and to recede even farther from the borders of South Africa. However, inadequate control measures in Mozambique since the mid-'70s have enabled tsetse flies to increase their range, and it is possible that they may re-establish themselves in many of their former areas of distribution. In general, tsetse flies avoid open country and prefer well-wooded areas. The various species do have environmental preferences, however, some living in forested areas and others in bushveld, and even within these areas they seek out locations of suitable humidity and other conditions.

PLATE 29

Family Hippoboscidae (Louse flies)
Luisvlieë

Despite their unusual appearance, these are true flies which have been structurally modified to better suit their parasitic way of life. Most adults are medium-sized and have a flattened body with a tough, leathery cuticle to make it difficult for the host to grasp or squash them. They have well-developed legs tipped with sharp, hooked claws as additional adaptations for clinging firmly to the host. Wings may be present or absent, depending on the species. Adult louse flies are ectoparasites which feed on the blood of birds or mammals.

No eggs are laid by members of this family. Instead, in much the same way as in tsetse flies (page 120), a single larva develops inside the body of the female and is born when fully grown. In most cases the larva then digs into the soil or finds a protected place, where it pupates. In some species the larva pupates on the body of the host or, in the case of some birds, in the nest material.

Although it does not occur in the KNP, probably the best known hippoboscid is the sheep-specific sheep ked, *Melophagus ovinus*. When hippoboscids occur in large numbers on a host they may cause such constant irritation and blood loss that the host loses condition and, in severe cases, dies. Generally, however, such animals have already been weakened by disease or other factors which predispose them to being heavily parasitized by hippoboscids.

Although little effort has been made to determine which species are present in the KNP, several species are known to be very common. Zumpt (1966) describes at least 25 species which, judging by their distribution and host range, are likely to be found in the KNP if suitable collecting efforts are made.

PLATE 30

Family Diopsidae (Stalk-eyed flies)
Steeloogvlieë

Stalk-eyed flies are among the most distinctive of all flies, and there is little chance that they will be mistaken for any other insect. Most have their eyes situated on slender extensions which project from the head. Adults are small to medium-sized, and have a slender body. The wings at rest are held back over the body. They appear to feed on oozing liquid from plants. The larvae have been found in decomposing vegetation, while some tunnel into living plants for nutrients.

Stalk-eyed flies generally occur near water on grasses and sedges in the warmer areas of Africa. Within the KNP they have been recorded, commonly but patchily distributed, in the riverine forests along the Sabie, Olifants and Luvuvhu rivers.

PLATE 30

Family Platystomatidae (Redheaded flies)
Rooikopvlieë

These are medium-sized to large flies, often with a metallic coloration. The adults generally feed at flowers, decaying fruit, faeces and even decomposing snails. The larvae have been found in various vegetable substances and also in legume root nodules.

A species which regularly attracts attention in the KNP is *Bromophila caffra*, a 13-mm-long fly of somewhat daunting appearance: it is dark metallic-blue in colour and has large red eyes. Despite its looks, it is completely innocuous and is generally found singly or in small groups resting on the leaves of twigs in a low tree. When disturbed, it flies off rather weakly. It depends on its bold warning colours to advertise its foul taste and discourage predators.

PLATE 30

Family Conopidae (Thick-headed flies)
Dikkopvlieë

The Conopidae are smallish to medium-sized flies which hold their wings out sideways when at rest. Most conopid adults strongly resemble wasps (page 135) in appearance. The base of the abdomen is often constricted as is common in many wasp families, and even the antennae are modified to resemble those of a wasp.

Adults deposit their eggs on wasps and bees. The larvae burrow into and live as parasites within the body of their host.

Although the family occurs throughout Africa, these flies are not often seen; when they are, it is generally in the riverine forests of the KNP.

PLATE 30

Family Piophilidae (Skipper flies)

This family is well known due to a single species, *Piophila casei*, which is distributed worldwide and has larvae which live on cheese, ham, bacon and other fatty, high-protein substances. The family contains few species, most of them small in size and dark in overall coloration.

The cream-coloured larvae of *Piophila* have the conspicuous habit of bending themselves into a circle, hooking their mouthparts on to the posterior extremity of the abdomen, pulling hard, and then snapping free with a jerk which sends them flying several centimetres into the air and well away from a predator or otherwise unfavourable position.

Piophilids are common at carcasses throughout the KNP and Africa. In 1978 a new species, subsequently named *Piophila megastigmata*, was discovered at carcasses in the Pafuri area of the KNP.

PLATE 30

ORDER TRICHOPTERA (Caddisflies)
Kokerjuffers

Few people besides entomologists know of the existence of caddisflies; these insects so closely resemble moths (page 124) that they are generally accepted as such. The adults have a typically moth-like shape and form, but can easily be distinguished by two features: firstly, whereas moths have their wings densely clothed in small overlapping scales, caddisfly adults have few scales, if any, instead having a layer of hairs; and secondly, moths generally have their mouthparts modified into a long, coiled-up proboscis whereas caddisflies have rudimentary or no mouthparts. Adult caddisflies tend to be medium-sized and dull in colour, mostly brown. They have long, thin antennae which are held straight in front of the head. They are nocturnal, and are often attracted to lights on summer evenings.

The larvae are elongate, resembling caterpillars. All live underwater in rivers and streams, where they form an important source of food for other aquatic insects and fish. Most caddisfly larvae build a hollow, cylindrical, protective case, in which they live. The case is made of grains of sand or pieces of vegetation which are chewed into the correct shape for building. The sand or other strengthening material is embedded in a base of silk produced by glands in the head of the larva. Different species use different materials or different sizes of the same material to build their larval cases. The shape of the case – whether straight, spiralled, even-sided, tapering or irregular – also varies, so that it is often possible to identify a species by its case. Most case-bearing larvae feed on vegetation or organic detritus. Some species have no protective cases; these larvae spin small silken nets between stones or vegetation to catch other insect larvae or tiny Crustacea, while other species rapidly dart after and catch their prey.

All larvae, whether case-building or not, pupate within a case which if not made in early larval life is built immediately prior to pupation. The pupal case is closed at both ends, but not entirely sealed, so that water still flows through freely, thus ensuring a plentiful supply of oxygen. When ready to emerge as an adult, the pupa makes its way out of the pupal case and moves to the edge or the surface of the water. Here the adult emerges and flies off.

Trichopterans are common along most unpolluted lakes and rivers in Africa, but are not often seen due to the aquatic lifestyle of the larvae and the nocturnal activity of the adults. In the KNP they are frequently seen at lights at night in all the camps situated near rivers.

ORDER LEPIDOPTERA (Moths and butterflies)
Motte en skoenlappers

This is probably the group of insects best known to most people, with many species admired for their bold and beautiful colours. It is a very large order and in terms of diversity is second only to the Coleoptera (page 95). With the exception of a few species which have wingless females, all adult moths and butterflies have two pairs of wings which are covered with tiny, overlapping scales. Most species also have a long proboscis – coiled up when not in use – for sucking juices from flowers, fruit, dung, mud and other sources.

Moths and butterflies make up an abundant and essential source of food for a wide range of animals, including other insects which parasitize or prey heavily on especially the larval stages of Lepidoptera. Many species of wasps and flies feed exclusively on caterpillars during certain stages of their life cycles. Humans also eat the larvae of some species, such as those of the mopane moth, *Gonimbrasia belina*, thousands of which may be collected when they reach abundance during summer.

A large number of species in the order, moths especially, are pests, destroying, disfiguring or contaminating various materials in their larval stages, when they feed voraciously. Many caterpillars will bore into fruit, others into wood and seeds, and some into stored products such as grain.

There are four suborders within the Lepidoptera. Only one of these is likely to be seen by visitors to the KNP.

SUBORDER DITRYSIA

More than 97 per cent of the known lepidopteran species are contained in this group, the most advanced of the four suborders. The females have two genital openings: a copulatory pore on the ventral surface of the eighth abdominal segment, and an opening on the ninth segment through which the eggs pass during laying.

PLATE 31

Family Psychidae (Bagworms)
Sakwurms

This is an unusual family in that the females are wingless. The females are often degenerate in other ways, too, such as in having reduced mouthparts, antennae or legs. The males, on the other hand, have a well-developed body and are capable of rapid flight. Their wings are sparsely covered with hairs and imperfect scales, but have no bright colours or other distinctive markings, thus appearing rather drab.

The larvae live in self-constructed silken bags attached by silken strands to twigs or branches. The bags are strengthened and camouflaged by the embedding of leaves, twigs, grass stalks and other fragments of vegetation in the silk pieces. When the larva outgrows its bag, a larger one is constructed, and it is often possible to distinguish the species from the form and material of the bag. When fully grown the larva pupates within its protective bag. After emerging, the adult males fly off in search of females, which remain within the bag. Having located a female, the male inserts his extensible abdomen into the bag to fertilize the female. The eggs are laid within the bag, after which the female dies. Adults generally live for only a few days.

Psychids are widespread in Africa in all the well-vegetated areas, with about 30 species recorded from the southern part of the continent. One of these, the wattle bagworm, *Kotochalia junodi*, is a pest on wattle foliage, but also feeds on the leaves of a wide range of acacia trees. Bagworms have been found in riverine forest, but are more common in open bushveld throughout the KNP.

PLATE 31

Family Tineidae (Clothes moths and their relatives)
Kleremotte, wolmotte

These are small, inconspicuous moths with fairly narrow wings and a proboscis which is either short or absent.

The larvae vary considerably in habits. Some feed on vegetation; others, unusual in that they have enzymes capable of digesting keratin, consume the horns of dead animals, feathers, woollen carpets, clothing and fur. Some larvae live in self-made bags of silk embedded with fragments of food substrate.

Three species of clothes moths – *Tinea pellionella*, *Tineola bisselliella* and *Trichophaga tapetziella* – occur commonly in Africa and have also reached pest status in many countries on other continents. *Ceratophaga vastella*, the most commonly seen species in the KNP, is widely distributed over Africa wherever horned animals are abundant. The adults of this species have creamy-yellow forewings and greyish hind wings. The larvae feed on and tunnel into the keratinous sheath surrounding the bony core of the horns of dead bovids. Several hundred larvae may be found in a pair of horns, each larva constructing one or more narrow 'towers' several centimetres in length and consisting of cemented faecal pellets. Partially extruded, empty pupal cases are often seen projecting from the ends of these towers after the adults have emerged.

Tineids are common in most areas of Africa. They have been found in all parts of the KNP.

PLATE 31

Family Zygaenidae (Burnets)

Small to medium-sized moths with narrow wings, burnets often have antennae which are thickened at the tips. The wings of many species are often brightly coloured with patterns of red, green or blue on a black background. These contrasting colours serve to warn potential predators that the moths are foul-tasting. In some species this bad taste – which is also toxic – is due to the cyanide compounds contained in the body of the larvae, the precursors of which are derived from the plants on which they feed. The toxic substances are retained in the body when the larvae become adults. Brightly coloured moths tend to be diurnal, whereas the dull species are usually nocturnal.

The larvae are short, squat and cylindrical in shape, and generally feed on the leaves of herbaceous plants. When fully grown they pupate above ground, usually on a twig, within a self-spun, silken cocoon.

The family is widespread over much of Africa, but appears not to be abundant; nearly 100 species occur in southern Africa. Several species occur in the KNP, where they prefer to remain close to riverine forest or other well-wooded areas.

PLATE 31

Family Pyralidae (Pyralids)

This is a large family of generally inconspicuous, drably coloured moths, ranging in size from small to large. The labial pulps are prominent and projected forwards or upwards as a beak-like structure from below the head. Adults have hearing organs at the base of the abdomen for reception of sound.

Larvae are found in a wide variety of situations; feeding within the stems, fruits or seedheads of plants, in silken galleries among mosses and leaves or in stored human food products.

Members of the subfamily Galleriinae are small, greyish moths. Some caterpillars live on the wax combs of bees and wasps, some on dried fruit, and others on underground roots of trees. The most well-known of these is the cosmopolitan wax moth *Galleria mellonella*, which wreaks havoc in the wax combs of honey-bees, frequently becoming an economic pest to bee farmers. *Achroia grisella* has similar habits but is less often a pest.

Cactoblastis cactorum is a smallish grey moth of the subfamily Phycitinae, and was introduced into South Africa in 1933 for the biological control of the prickly pear *Opuntia megacantha*. In 1988 in a joint venture between the National Parks Board and the Department of Agriculture and Water Supply, eggs of this species were placed on exotic *Opuntia ficus-indica* which was a problem species present in a large area around Skukuza. The eggs hatched successfully, caterpillars ate their way into the prickly pear cladodes and a small thriving colony was established. Already this moth shows considerable potential in assisting control of *Opuntia* originally brought into the KNP by humans earlier this century. Despite extensive tests previously made by state entomologists to ensure that this species does not utilize indigenous plants, a recent 'scare' arose when a related species was discovered feeding on and severely damaging *Euphorbia* trees around Olifants and Skukuza camps. However, this appears to be a naturally-occurring species and there is no cause for concern.

Another familiar member of this family is the jumping bean moth *Emporia melanobasis*. This moth feeds and lives within the seeds of the tamboti tree *Spirostachys africana*, and when these seeds are dropped in September or October of each year the activity of the beans 'jumping' up – several centimetres at times – always attracts the attention of tourists. This activity appears to be mostly when the seeds are lying in the sun and the tiny caterpillars become overheated. They curl up, hold on to the posterior abdomen with their mouthparts, exert tension and suddenly release so that they are flicked up with the seeds, hopefully moving closer to a shady position.

PLATE 31

Family Pterophoridae (Plume moths)
Veermotte

This is a family of small to medium-sized, delicately built moths which tend to be plain or inconspicuously coloured, but are nevertheless distinctive in appearance. The narrow forewings are split into two (sometimes three or four) distinct divisions by deep clefts. The hind wings, also narrow, are usually split into three separate, elongate divisions. The legs are long, slender and frail, and have spiky projections (tibial spurs) near their extremities. When resting, many species raise some of their legs like antennae.

The larvae generally feed on the flowers and leaves of the plant order Compositae, but often feed on other plants, sometimes internally in the stems or seeds.

Very dainty and appealing insects, plume moths occur in many parts of Africa, although not much is known about their habits. They have been recorded from many localities in the KNP.

PLATE 32

Family Geometridae (Loopers or earth measurers)
Landmeters

This large family contains mostly medium-sized moths which have large, rounded wings and a slender body. Most species are inconspicuous in colour, displaying patchwork patterns or wavy blotches that blend with a background of bark, rock or soil. They do not have a strong flight, and when resting the wings are usually held horizontally against the substrate. The larvae are easily recognizable: they are slender and, besides the three pairs of true thoracic legs, have only two pairs of fleshy prolegs on the abdomen, unlike most caterpillars which have five pairs. These two pairs are situated near the hind end of the body on the sixth and 10th abdominal segments, leaving a considerable length of the body unsupported.

Their characteristic walking action has given rise to their common name. When moving, the caterpillar brings the hind end of the body forwards to the thorax, causing the central part of the body to twist into a C-shaped loop above. The thoracic legs then release their grip on the substrate, the anterior part of the body is extended forwards until the body is straight, and the hind end is once again brought forwards by looping the central part of the body.

Most larvae feed in the open on herbaceous vegetation, but their coloration and knob-like outgrowths usually render them inconspicuous among twigs or other parts of the background. Many have the habit of grasping a twig with the abdominal prolegs, then holding the body stiffly at an angle to the twig, remaining motionless in this position for long periods, very closely resembling a broken branchlet. When fully grown the larva pupates either in the soil, or in a bedraggled cocoon between leaves, or in a concealed position such as in the cracks of deep bark.

Geometrids are widespread and numerous over all plant-supporting areas of Africa, with more than 1 000 species having been recorded in southern Africa. They occur in all vegetation zones throughout the KNP.

PLATE 32

Family Saturniidae (Emperor moths)
Pouoë

These medium to very large moths generally have the mouthparts either reduced or absent. Most saturniids may be recognized by their beautifully and brightly coloured wings, each with its characteristic 'eye spot'. When disturbed, the moth may suddenly jerk the forewings up to expose the larger eye spots on the hind wings, which often has a sufficiently disconcerting effect to frighten off a potential predator. The body, legs and wings are covered with a dense fur. Males tend to be smaller than females.

The adults of most species are nocturnal and have a preference for open bushveld, with only a few species occurring in dense forests. The females usually lay their eggs in clusters on the food plant of the larvae. Young larvae are generally darker and hairier in appearance than the later stages; older larvae often have rows of spiky projections or tubercles adorned with spines over most of the body. When fully grown, the larvae pupate in the soil or among leaf-litter, with some species constructing silk cocoons.

The larval stages of *Gonimbrasia belina*, known as mopane worms, are often very abundant in summer in those northern areas of the KNP dominated by mopane trees (*Colophospermum mopane*), where tens of thousands of these worms can be seen. Regarded by some people as nutritious delicacies, the large caterpillars are collected in bags, squashed, spread out to dry in the sun, and finally added to stews or broths. The larvae are not restricted to mopane trees but feed on a variety of other shrubs or trees where *Colophospermum mopane* is not present. Another exceptionally beautiful species occasionally seen in the KNP is *Argema mimosae*, the adults of which are large, delicately green in colour, and have hind wings with long, twisted tails. The larvae are generally found feeding on marula leaves.

Many species of emperor moths occur in Africa. The KNP has a diverse range of saturniid inhabitants, most abundant in summer.

PLATE 32

Family Sphingidae (Hawk moths or sphinx moths)
Pylsterte

These small to large moths are often attractively coloured. The adult's body is very thick but streamlined, being somewhat blunt at the anterior end but, in most species, tapering to a fairly sharp point posteriorly. The antennae thicken gradually towards the tips, where they are thin and bent back, hook-like. Most species have a long, coiled proboscis and well-developed eyes. The wings are characteristically adapted for rapid flight, the forewings being long and narrow and the hind wings small.

Most adults are nocturnal or crepuscular, although a few species are active by day. They have a fast, darting flight, occasionally hovering in front of flowers for a sip of nectar.

The cosmopolitan death's head moth, *Acherontia atropos*, which is recognizable by the skull-like pattern on its thorax, is unusual in that it raids beehives, feeding on the honey. Most species of hawk moths are important as pollinators of plants.

The larvae feed on leaves. They are generally smooth, lacking hairs, and have a characteristic large, fleshy spike or hook-like projection near the hind end of the body. Many larvae also have large spots resembling eyes near the anterior end, giving them the appearance of a snake, which serves to frighten off predators.

Pupation usually takes place in a small hole tunnelled into the soil, or in a loosely constructed cocoon in leaf-litter above the soil. The pupae are recognizable by the fact that part of the proboscis is looped free of the main pupal body, much like a cup handle.

Hawk moths are well represented in Africa, with about 200 species occurring in the southern part of the continent. Many of these are present in the KNP.

PLATE 33 | **Family Thaumetopoeidae** (Processionary worms)

The moths in this small family are medium-sized, have a stout body, well-developed, rounded wings which are often attractively patterned, and lack mouthparts.

The family is best known for the unusual habits of the larvae, the hairs of which cause severe irritation if brought into contact with the skin. Found in groups of up to 600, but usually much fewer, the hairy caterpillars are gregarious and often found tightly clustered on the lower parts of a tree trunk or shrub, usually that of their food plant. When moving on the ground in search of a new feeding site, the group strings itself out in a long column, each caterpillar touching the one in front. A line of silk is deposited as the column marches on but this does not appear to be significant in maintaining direction, head to tail contact being the essential factor. If the column is broken for some reason, the severed section of the column will stop abruptly and the foremost larva will move the anterior part of its body in a random search for the larvae ahead. Eventually the front caterpillar of the broken part of the column will assume leadership and begin walking again, and the procession will follow.

When ready to pupate the larvae cluster together and a silken envelope is spun over the entire group. Each caterpillar then constructs its own cocoon under this protective covering, so that eventually a compact mass of cocoons is formed, consisting of silk encrusted with hairs from the skins of the larvae and impenetrable to most predators.

The adults are short-lived, with their sole purpose being to disperse and reproduce.

Processionary moths are widespread over Africa, with about nine species occurring in the southern part of the continent. In the KNP they have been recorded only south of the Olifants River. *Anaphe reticulata* appears to be the most commonly encountered species. The adult processionary moths have highly attractive, cream-coloured forewings patterned with black striations.

PLATE 33 | **Family Noctuidae** (Owl moths)
Uilmotte

This is the largest family in the Lepidoptera. Most are medium-sized insects with a well-developed proboscis, and have wings patterned in shades of brown and grey, the hind wings generally being lighter in colour. Most species are nocturnal both in the larval and adult stages, although a few, brightly coloured, species are active by day. The larvae tend to be smooth and hairless, and feed on the foliage or stems of herbaceous plants. When fully grown, they usually pupate in the soil. However, many exceptions to this general description exist.

Although the adults of most species feed on the nectar of flowers, some species, like *Arcyophora longivalvis*, *A. patricula* and *A. zanderi* in Africa, feed on the lachrymal secretions (tears) around the eyes of various mammals. Others, such as species in the genera *Othreis*, *Calpe*, *Serodes* and *Dugari*, have a strengthened, saw-like proboscis which is used to pierce fruit and suck the juices. *Calpe eustrigata* in Asia uses its proboscis to pierce the skin of mammals to suck blood.

Many pests of agricultural products are members of this family, including the bollworm (*Heliothis armigera*), the cutworm (*Agrotis segetum*), the tomato worm (*Spodoptera littoralis*), the lesser army worm (*Spodoptera exigua*), the maize stalk borer (*Busseola fusca*), the red bollworm (*Diparopsis castanea*) and the African army worm (*Spodoptera exempta*).

The larvae of the African army worm, found in Africa, Australia and parts of Asia, and recognizable by their green or pink and green coloration, are usually present in low numbers. In some years, however, when environmental conditions become exceptionally favourable (high temperatures, high humidity, an abundance of succulent young grasses and low numbers of natural predators and parasites), a massive increase in the numbers of larvae may occur. In a manner similar to locusts, this increase in density results in the caterpillars changing colour and behaviour, and entering the 'gregarious phase'. The caterpillars become velvety-black above and yellowy-green below, and swarm over the countryside feeding on members of the grass family, including maize, wheat and sugar cane, but also on cotton and potatoes. When fully grown the larvae pupate in the soil, and the adults emerge one to four weeks later, depending on the temperature. After copulation, the females lay their eggs in batches of 200-300 on the underside of the leaves of the food plant, and the cycle resumes. Environmental conditions (adverse climatic conditions or a buildup of predators and parasites) sooner or later bring about a sudden collapse in population numbers, after which the army worms revert to the 'solitary phase'.

Noctuids are very common and widespread over the African continent, with about 1 700 species occurring in southern Africa. Large numbers are found in all the vegetation zones of the KNP, many being attracted to lights in the evenings. Periodic outbreaks of army worms occur in the KNP, with the focus appearing to be the Bangu/Gudzane windmill areas in the grassy plains of the Central District.

PLATE 34

Family Ctenuchidae (Handmaidens)

Often mistakenly thought of as members of the family Zygaenidae (page 125) because of their similar appearance, the ctenuchids form a small but distinct group of moths. The adults have narrow forewings and small hind wings, both pairs generally being purplish or black in colour with several transparent areas, although these may be coloured in some species. The rather stout body tends to be blackish, often having transverse bands in various colours. Most species are diurnal, and most are slow in flight. The larvae are short and squat, and have tufts of hair along the body. Most feed on grasses or lichens, and occasionally also on shrubs and trees. When fully grown, they pupate in a cocoon matted with setae from the larval skin.

Although apparently never abundant, ctenuchids are widespread in the warmer, vegetated areas of Africa, being most common in the tropics. About 30 species are present in southern Africa; several of these occur throughout the KNP.

PLATE 34

Family Arctiidae (Tiger moths and woolly bears)
Tiermotte

This large family consists mostly of medium-sized, stout-bodied, broad-winged moths, although there are exceptions. The wings of many species are white, yellow or black, spotted or banded with contrasting hues. Most are nocturnal, although some species are active by day.

The larvae generally feed on grasses, herbaceous plants or lichens, and are often covered with a dense coat of long hairs – hence the common name – as a repellant against predators. The hairs are usually incorporated into the thin, silk cocoon during pupation, again to discourage predation.

The Arctiidae is a widespread and common family in Africa, with about 300 species occurring within southern Africa. Many of the species are found widely distributed over the KNP, where they are often attracted to lights at night.

PLATE 35

Family Papilionidae (Swallowtails)
Swaelsterte

With their large, well-formed wings, this family includes some of the most beautiful butterflies found anywhere on earth. The wings are often black, patterned in shades of white, yellow, red, green or blue. The antennae are knobbed – as are those of all butterflies – and the thorax and abdomen tend to be slender.

The adults are active by day, and are often seen poised on flowers sipping nectar with the well-developed proboscis; sometimes they sit on the mud-lined banks of pools, where they suck at the moisture. Their flight is fairly rapid and bouncy.

Most larvae have a smooth cuticle, although some species do have hairs. As an aid to defence the larvae have repugnatorial glands which can be extruded through a slit in the prothorax. This protrusible structure, known as the osmeterium, consists of a two-lobed sac that gives off a bad odour, and is used only when the larva is threatened or disturbed. The swallowtail pupae are generally attached to a twig by means of silk strands which are hooked to the posterior and middle of the body. A few species pupate in a silken web roughly constructed between leaves.

Although it is not a particularly large family, the Papilionidae nevertheless occur commonly and are widespread in the warmer, well-vegetated areas of Africa. One species, *Princeps demodocus* – known as the citrus swallowtail or Christmas butterfly – is a pest during its larval stages on the foliage of citrus trees. Ten species have been collected in the KNP. Of these only three are widely distributed and common, the others being more patchy in their range. One of the three, the beautiful and rare mocker swallowtail, *Princeps dardanus cenea*, has four different female forms, all of which mimic the distasteful butterflies of the family Danaidae (page 134).

PLATE 36

Family Hesperiidae (Skippers)

This is a large and widespread family with the majority of species being small to medium in size. Most are drably patterned in white and shades of brown although some tropical species are brightly coloured and grow to a large size. The antennae usually thicken gradually towards the distal extremities, but are sometimes thin and recurved as hooks at the tips. The body is stout, compact and hairy. The thorax is large and crammed with the muscles necessary to power the fast, erratic bursts of flight from which these butterflies derive their popular name. They are active by day or at dusk, when they are often seen feeding at flowers or resting on twigs, or occasionally on the ground or against rocks. When stationary, the wings are held either vertically above the body or horizontally, parallel to the substrate.

The larvae also tend to be rather stout and are tapered towards either end. Many species feed on grass blades, some on shrubs, and others on banana trees. Most remain in the open on the food plant, although they are inconspicuous because of their camouflage coloration; some live in webs or between leaves partially held together by silk.

When fully grown the larvae either spin a rough cocoon in leaf-litter and pupate inside it, or attach themselves to a twig by means of a silk support hooked onto the hind end of the body, and often also a silk loop which encircles the insect at mid-body.

This family is well represented in Africa, distributed wherever vegetation occurs. Forty-one species have been collected in the KNP. These are illustrated and described in 'Butterflies of the Kruger National Park' by J Kloppers and G van Son (1978).

PLATE 36

Family Pieridae (Whites)
Witjies

Rare indeed are the sites in Africa where a diverse range of these attractive butterflies do not abound. Most are medium-sized, though small and large species are also common. The body colour tends to be white or yellow with black patterns, the forewings often having the tips shaded in red. The antennae are distinctly knobbed and the body is usually slender. Adults are diurnal and occur in open bushveld or in dense forests, often feeding at flowers, sipping moisture at mud pools or elsewhere, or resting quietly on vegetation. They generally have a fairly straight, somewhat bouncy flight of medium velocity.

The larvae are elongate, sometimes hairy but generally smooth-skinned, and do not have any fleshy protruberances, osmeteria, or other outgrowths projecting from the body. Most feed on plants belonging to the families Leguminosae and Capparidaceae. The pupa is usually attached to a twig by silk strands hooked to the posterior end of the body, and often also by a supporting girdle of silk looped around the insect at mid-body.

Pierids are widespread and common throughout the KNP, being especially abundant in autumn and spring. Riverine forests with open areas overgrown with low herbaceous plants often have swarms of these butterflies flitting around. Thirty-seven species have been recorded in the KNP, most of them being rather common. More details on their distribution and seasonal abundance are available in the well-illustrated guidebook 'Butterflies of the Kruger National Park' by J Kloppers and G van Son (1978).

PLATE 36

Family Lycaenidae (Hairstreaks, blues and coppers)
Bloutjies en kopertjies

This is a large family of small to medium-sized butterflies, the wings of which are generally coloured in shades of brown, blue, orange or copper on the upper surface and spotted or streaked below. The hind wings often have attractive, fragile, tail-like extensions posteriorly. The adults are diurnal. Most species fly close to the ground at medium velocity, although some are capable of very rapid flight.

The larvae differ considerably in habits and appearance. Some species are carnivorous, crawling on vegetation to prey on aphids, coccids and other homopterans. Others have a gland or glands on the abdomen which produce a sweet secretion; ants stimulate these glands by stroking the appropriate segment of the caterpillar with their antennae, and then lap up the drop of liquid which oozes out. Some caterpillars live on trees and shrubs, presumably under the indirect protection of the resident ants which chase off parasites and other enemies; others enter the nests of ants – either by themselves or they are carried in by their hosts – where they feed on the larvae, pupae and cocoon material of the ants. The ants endure this marauding for the sake of the sweet secretions provided by the caterpillar.

The family is well represented in most areas of Africa, there being approximately 1 400 recorded species in Africa. Lycaenids are present throughout the KNP, most abundantly in the open bushveld areas near small streams or pans. Sixty-one species have been collected in the Park. These are illustrated, and their distribution and seasonal abundance described, in the guidebook 'Butterflies of the Kruger National Park' by J Kloppers and G van Son (1978).

PLATE 37

Family Nymphalidae (Brush-footed butterflies)
Borselpootskoenlappers

All the butterflies in this group have reduced forelegs which can no longer be used for walking or clinging. The legs are often clothed in long setae, hence the popular name. Sexual dimorphism, where the sexes differ in appearance, is a common feature of this family and often striking. Nymphalid butterflies tend to be medium or large in size and are generally patterned in shades of brown or black, although several brilliant and beautifully coloured species occur. The adults are active by day, often being seen sipping moisture from the edges of pools or visiting flowers. The larvae are cylindrical in shape, often having spines or fleshy protruberances on the body. The pupae are usually found as naked individuals (having no cocoon), hanging from vegetation by a point of attachment at the posterior end of the body.

Nymphalids are widely distributed over the greater part of Africa. Many species are common in the KNP.

PLATE 37

Family Charaxidae (Swifts, charaxes)
Dubbelsterte

This family of medium to large butterflies is easily recognizable by the characteristically robust body and particularly stout thorax. Most species display cryptic coloration on the lower surface of the wings for camouflage, while the upper surface is bold and generally brightly coloured. Short, tail-like extensions are present posteriorly on each hind wing. The large thorax houses the powerful musculature necessary for the fast, erratic flight. The larvae are usually smooth, with a shield-like head and fleshy projections on the head and hind end of the body.

Most species confine themselves to forest or woodland, where they frequently can be seen darting around tree tops, feeding at wounds on trees where nutrient fluids exude, or on fresh elephant dung or even animal carcasses where they sip available fluids.

Charaxes butterflies are common inhabitants of natural woodland in Africa. A wide range of species is present in the KNP.

PLATE 38

Family Danaidae (Milkweed butterflies, monarchs)
Melkbosskoenlappers

The butterflies in this family have a relatively thin body but large wings, generally coloured in shades of brown, black and white. They are popularly known as milkweed butterflies as most of the larvae feed on members of the plant family Asclepiadaceae or close relatives, deriving toxic substances (cardiac glycosides) from the leaves. The substances are retained within the body so that even in the adult stage these insects are distasteful or toxic to predators. Birds, lizards and other animals normally feeding on butterflies quickly learn to associate the bold colours and patterns of the Danaidae with their foul taste, and so avoid them. Several unrelated butterflies mimic the colours and body shape of the Danaidae in order to similarly escape predation. Adult Danaidae are diurnal. They have a fairly slow and erratic flight pattern, but can be very effective in dodging to escape butterfly collectors.

A most interesting attribute of the Danaidae is the presence of 'hair-pencils' posteriorly in the males. These are brush-like structures normally concealed within the body, but extruded during courtship when they are used to brush a pheromone onto the antennae of the female. The wing glands on the hind wings of the males, most obvious in the African monarch butterfly (*Danaus chrysippus*) as prominent black spots, is actually a pouch which holds innumerable small scales; the hair-pencils are occasionally rubbed into this wing gland, which is presumed to have a complementary function to the pheromone already present on the hair-pencils. The females usually lay their eggs singly on plants of the Asclepiadaceae, where the boldly patterned larvae – which have several thread-like projections on their backs – feed until fully grown and pupate by suspending themselves from vegetation using small hooks at the posterior end of the body.

Although this is a small family, some of its species are very widespread, the African monarch butterfly being especially common over most parts of the continent as well as parts of Europe and Asia. Five species occur within the KNP, but only the African monarch is common and widespread, the others being localized.

PLATE 38

Family Acraeidae (Reds)
Rooitjies

Most of these generally medium-sized butterflies are reddish, spotted or streaked with black, and often have the forewings more or less transparent. However, several species are black with creamy or brown patterns. The body is long and thin with an unusually tough cuticle, allowing the insect to be roughly handled without suffering damage. When held or otherwise threatened, acraeids are capable of giving off an orange-coloured repellant fluid which is toxic or distasteful to predators; they are avoided by many insect-feeding birds and other animals which learn to associate their characteristic shape and colours with distastefulness. The chemicals responsible for this foul taste are derived from the food plants eaten by the larvae, but it is possible that some species actively synthesize repellant substances. The adults are active by day, when they can be seen flying slowly through open areas in search of low flowering shrubs or herbs.

The females lay their eggs singly or in clusters on the food plant. The larvae are cylindrical in shape, with many spiny projections on the body. As is the case with many other butterflies, the pupae are not covered with a cocoon but dangle loosely from vegetation by means of small hooks at the hind end of the body.

The Acraeidae is essentially an African group of butterflies, and is widespread and common in both savanna and forest. Seventeen species have been recorded in the KNP. Details of these are given in the guidebook 'Butterflies of the Kruger National Park' by J Kloppers and G van Son (1978).

PLATE 38

Family Satyridae (Browns)

These butterflies are generally medium-sized although several large species also occur. They have large wings relative to the slender body; these are generally coloured brown, often with 'eye spots' on the upperside of both pairs, but cryptically coloured on the underside. The adults tend to have a slow flight pattern and rarely fly far before settling on soil, leaf-litter or vegetation, where they are difficult to see due to their excellent camouflage coloration. The body of the larva may be smooth or covered with fine hairs. Most species feed on grass. When fully grown it attaches itself, by the hooks at the hind end of the body, to vegetation or stones, or lies unattached among fallen vegetation.

More than 250 species occur in Africa, in forests, savanna, and open grassland. Ten species have been recorded in the KNP, their distribution and abundance described in 'Butterflies of the Kruger National Park' by J Kloppers and G van Son (1978).

ORDER HYMENOPTERA (Wasps, bees, ants and their relatives)
Perdebye, bye en miere

This is the third largest order of insects, containing well over 100 000 described species. In terms of behaviour it is the most highly developed order, and it contains many social species. The winged members of this group have two pairs of membranous wings (hence the name of the order), the hind pair being the smaller and connected to the front pair by a series of small hooks. The first abdominal segment is fused to the thorax, and the second segment (sometimes also the third) is usually narrowly constricted to form a waist-like petiole or pedicel. The body length of these insects varies from less than 1 mm to more than 40 mm. The females of many species have their ovipositors modified into a sting with accompanying poison glands.

The order contains numerous parasitic species which help to control, among other things, lepidopterous pests of agricultural crops. Many hymenopterans prey exclusively on lepidopterous caterpillars, while others parasitize aphids and scale insects. This not only renders them beneficial to man but also means they play a large role in helping to maintain a natural balance in the ecological web. Many species are highly important as pollinators of plants, so essential that a number of plant species would not survive if it were not for these insects; in addition, honeybees produce a vast amount of honey for human consumption every year.

Two distinct suborders are recognized, based on external appearance. The Symphyta include the more primitive families, whereas the Apocrita include the wasps, bees and ants.

SUBORDER SYMPHYTA

These insects are easily recognizable by the broad junction between the first and second abdominal segments, distinct from the narrow, waist-like constriction of the Apocrita (see below). The larvae generally have thoracic and abdominal legs, and in many cases superficially resemble lepidopterous caterpillars.

PLATE 39 | **Family Tenthredinidae** (Sawflies)
Bladwespe

This is the largest family within the Symphyta, and the only one discussed here. The adults tend to be medium-sized, and have a wasp-like appearance apart from the broad area of attachment between the abdomen and thorax, typical of this suborder. The adults are generally inconspicuous in appearance, often having an orange or brownish colour, and occur in a wide range of vegetation types. Some species prey on other insects. The larvae are similar in appearance to the caterpillars of butterflies and moths, but differ in habits between the species. Most feed in the open on leaves but some tunnel inside soft stems, fruit or leaves. Whereas the larvae of the Lepidoptera have five or fewer pairs of fleshy abdominal legs, those in this family generally have six to eight pairs.

The common name is derived from the females, which have a saw-like ovipositor armed with serrations for making incisions in vegetation to conceal their eggs. Those which lay their eggs in hard wood have large teeth on the ovipositor, while those laying in soft wood have small serrations, and some species which only make a slit in leaves have hardly visible teeth.

Sawflies are common in most well-vegetated areas of Africa. Although poorly collected and documented in the KNP, they are certain to occur throughout its length.

SUBORDER APOCRITA

The first abdominal segment of insects in this suborder is fused to the thorax and followed by one or two segments which are constricted into a narrow, waist-like petiole. The larvae generally have no legs. The suborder is often subdivided into two divisions, namely the Parasitica (Terebrantia), which are parasites of other insects and in which the ovipositor retains its function of egg-laying, and the Aculeata, which embrace the hymenopterans that have the ovipositor modified as a sting. There are many exceptions to these two divisions, however.

PLATE 39 | **Family Ichneumonidae**

A very large family of small to large wasps, most of the members have a slender, elongate petiole and a dark pterostigma on the leading edge of each forewing. Many species are brightly coloured in uniform or combined shades of red, yellow, black and other hues. This is a highly important family as all the members are parasites on other insects or arthropods, although the caterpillars of moths and butterflies are by far the most frequently parasitized. Female ichneumonids tend to have a long, needle-like ovipositor which is jabbed into the body of the host insect prior to the eggs being passed through this channel. Many species use the ovipositor to puncture or drill through wood to reach beetle larvae or other immature insects, which they then parasitize. Most ichneumonid species have endoparasitic larvae which live within the body of the host, but many species are ectoparasites.

When fully grown the larvae emerge from the host and each spins its own silken cocoon, within which it pupates. Some of these cocoons are extraordinarily attractive, oval-shaped balls, variously patterned, and are suspended from vegetation by a thin strand of silk.

Ichneumonids are abundant and common throughout the vegetated areas of Africa. A large range of species occurs throughout the KNP.

PLATE 39

Superfamily Chalcidoidea (Chalcid wasps)

This exceedingly diverse assemblage of tiny (between 1 mm and 4 mm in length), mainly parasitic wasps embraces some 20 families. The wings of all species have a highly reduced venation, consisting essentially of a single vein adjoining the anterior or leading margin of each wing. Some species frequently form galls on the leaves, stems or even flowers of many plants, while other species deposit their eggs within the bodies of other insects, spiders and apparently even ticks, using the thin, syringe-like ovipositor for this purpose.

The fig wasps of the family Agaonidae deserve mention for their importance as pollinators of fig trees. All species of figs, including the enormous sycamore fig (*Ficus sycomorus*) that is so common in the KNP, have their tiny flowers enclosed within the 'fruit' and therefore out of reach of wind or other insects which could bring about pollination. Each species of tree has a limited number of fig wasp species which are adapted to enter the fig fruit to pollinate and reproduce; this is perhaps one of the best examples of co-evolution among plants and animals, for neither would be able to survive without the other.

The female fig wasp laboriously struggles its way through the closely overlapping miniature bracts or 'leaflets' of a young fig fruit, in the process losing its wings and antennae, and sometimes becoming trapped and dying.

Once inside the fig, the wasp locates the small female flowers and, using pollen from a special pouch on her body, deliberately pollinates the stigmas. During her wanderings inside the fruit she occasionally inserts her ovipositor into the ovules of some of the flowers to lay eggs, simultaneously injecting a secretion which causes the ovules to form galls. Within these galls the wasp larvae feed and grow, the male wasps growing and maturing faster than the females. When the males emerge they set about finding galls containing female wasps, puncture these galls and impregnate the females. This done, the males tunnel holes through the side of the fig to allow in oxygen from the outside, decreasing the high carbon dioxide content inside, a process which is essential for the proper development of the fig. The decreased carbon dioxide concentration activates the female wasps, which exit their galls and walk around collecting pollen which is stored in the body pouches. They then exit the fig through the tunnels made by the wingless males, now dead, and fly off to another young fruit to continue the life cycle.

Often when a ripe fig is broken open, numerous tiny black wasps are found inside, many of which have an exceptionally long, thin ovipositor projecting backwards from the end of the abdomen. These wasps, which belong to a different family, crawl out through the exit holes of the agaonid wasps, find another fig at a suitable stage of growth and push the long ovipositor through the fruit to lay eggs inside the developing agaonid larvae.

Members of the Encyrtidae and Aphelinidae play an important role in biocontrol of many agricultural pests by laying their eggs inside and parasitizing such insects as citrus red scale, pernicious scale and several other scale insects, as well as citrus psylla and various aphids.

The Pteromalidae is a large family with several well-known members, one of these being the widely distributed *Nasonia vitripennis*, a fairly common parasite of blowfly pupae at carcasses and other fly species in the KNP. Many other species of pteromalid, and also members of the Chalcididae, are important parasites of caterpillars and a range of other insects, helping to control their numbers and so maintaining balance in the ecosystem.

PLATE 40

Family Chrysididae (Cuckoo wasps)
Koekoekwespe

Very attractive and distinctive, cuckoo wasps generally have a brilliant green or blue, metallic coloration and an extraordinarily tough cuticle which is typically pitted with numerous small indentations. The abdomen is short, compact and convex above, and often strongly concave below. When caught or otherwise threatened the insect bends its body so that the head and ventral surface of the thorax fit into the concavity of the abdomen, with the result that the curled-up body resembles a ball with only the wings jutting out.

Chrysidids are most often seen on hot days, flying around or inspecting the nests of other Hymenoptera. The vast majority are parasitic on other insects and generally lay their eggs within the nests of solitary bees and wasps, a habit which has given rise to their common name. Pupation takes place within the hosts' nest, out of which the adults chew their way after emerging from the pupae.

Cuckoo wasps are fairly common and widespread throughout the warmer areas of Africa. In the KNP they are common around Pafuri.

PLATE 40

Family Mutillidae (Velvet ants)
Fluweelmiere

Small to medium-sized insects with a fine coating of hairs giving them a velvety appearance, mutillids are black or reddish-brown, often patterned on the abdomen with white bands or spots. The cuticle is extraordinarily tough and densely pitted with shallow indentations. The females, which are capable of inflicting a very painful sting, are wingless and are often seen walking along the ground or on rocks or walls in search of the nests of other hymenopterans, which they parasitize. The males generally have wings, and they often grasp the female and carry her off, copulating in flight.

Females deposit their eggs in the nests of other wasps and bees. The larvae feed on the host larvae or pupae. A few species parasitize

the larvae of beetles, moths and flies, including tsetse flies. Pupation takes place within the nest of the hosts, and adults chew their way out.
   Velvet ants are common throughout the warmer areas of Africa. They occur throughout the KNP, most abundantly at Pafuri.

PLATE 40

Family Pompilidae (Spider-hunting wasps)
Spinnekopjagters

This is a large family of small to very large wasps which are usually dark in colour and have well-developed legs, the hind pair being especially long. The antennnae are often long and twirled back in a circle. All are parasitic on various species of spiders.
   The females can be seen running around on the soil, frequently flicking or twitching their wings and antennae, searching through undergrowth or debris for suitable spiders to parasitize. When found, the prey is stung into insensibility (although some species kill the prey when stinging it) and either dragged or flown back to the nest, which generally consists of a simple hole in the ground, or a crevice or crack in wood or a boulder. A single egg is laid on the paralyzed victim , after which the hole is sealed with soil and vegetation. The female then flies off to repeat the process at another site. Inside the nest the hatched larva feeds on the spider until fully grown. It then pupates, and after emergence the adult tunnels its way upwards to the surface. A few species are ectoparasites on living spiders.
   Pompilids are common in most parts of Africa. In the KNP they reach maximum abundance in the Pafuri area.

PLATE 40

Family Eumenidae (Potter wasps or mason wasps)
Pleisterperdebye

This is a large family of medium-sized to large, solitary wasps. They usually have a bluish-black body, often marked with yellow or ochre.
   There is a wide range of nest structures, females either tunnelling into the soil or building mud nests against logs, twigs or stones. These mud nests may be multi-chambered or, as in the case of those wasps which sculpt highly attractive vase-shaped nests, contain a single cell or chamber. An egg is laid in each chamber and is provisioned with one or more lepidopterous caterpillars which have been stung into insensibility by the female. The chamber is then sealed off with mud. When hatched the wasp larva feeds on its supply of food. It spends the pupal stage in the same protective chamber, and the adult chews its way out after having emerged from the pupal case.
   Potter and mason wasps are common in most of the warmer areas of Africa. In the KNP they reach maximum abundance and diversity in the Pafuri area.

PLATE 41

Family Vespidae (Social wasps)

Brownish-coloured with a typical wasp appearance, these social insects live in communities which may be very large. Fertilized queens, unfertilized workers and males normally comprise a community, although some species lack workers, their function being taken over by immature females. The vast majority build nests composed of a single layer or multiple layers of combs, similar in appearance to those of honeybees, which are suspended from overhanging ledges, rocks, branches or roofs, although some conceal their nests in underground holes. Finely chewed plant material mixed with saliva is used for construction, giving the nest a papery consistency.

A community is started when a single fertilized female finds a suitable location for building a nest. She constructs a small comb made up of only a few cells, and lays an egg in each. She feeds the hatched larvae bits of chewed caterpillar and finally, when the larvae are fully grown and ready to pupate, seals off each cell. The adult males do not work, their sole function being to copulate. Females may be workers or queens. Unfertilized eggs result in male offspring, whereas fertilized eggs produce females.

Social wasps are common in most parts of Africa. They occur throughout the KNP.

PLATE 41

Family Sphecidae

With only a few exceptions, the members of this huge family lead solitary lives. Both sexes have wings and a typically wasp-like appearance, and tend to be medium to large in size. Most species dig into the soil to form an underground nest, although a few nest in crevices or hollow twigs, and some construct above-ground chambers or cells using mud. Each egg chamber is provided with an egg and stocked with insects and other arthropods which have been paralyzed or killed by a sting from the female wasp (the toxin of which also prevents the victim from rotting for several weeks), after which the entrance is sealed. Most species show no interest in the further development of the young, although a few exceptions do occur, such as in the genus *Bembix* where the adults build an open nest in sand and constantly feed their young with flies.

The Sphecidae is divided into several subfamilies, each being fairly distinctive in appearance and habits. The family as a whole can be considered to be beneficial to humans as they prey on house flies, blowflies and horse flies, and other arthropods of medical and veterinary importance.

Sphecids are common and widely distributed over most of Africa. Many species occur in the KNP, especially in the Pafuri area.

PLATE 41

Family Megachilidae (Leafcutter bees, carder bees and their relatives)
Blaarsnybye

Most of the members of this large family are medium-sized, brown or blackish, much like a honeybee in shape, and – except for the parasitic species – have a dense pad of hairs on the underside of the abdomen where pollen grains are stored. Most construct nests in hollow twigs, tunnels or cavities in the soil, holes in logs, walls or rocks, and even in key-holes and snailshells.

The female builds a series of cells using mud, resin or other plant material or, in the case of the leafcutter bee, bits of leaves. A single egg is laid in each cell and provided with a tacky but nourishing mixture of pollen and honey. The parasitic species do not bother to construct nests, but instead deposit their eggs in the nests of other megachilid bees. Their larvae then kill either the egg or the larva of the legitimate occupant and feed on its food supply.

Megachilid bees are numerous and common over much of Africa. Many species occur and are widespread in the KNP.

PLATE 42

Family Xylocopidae (Carpenter bees)
Houtbye

Although this is a comparatively small family, some of its species are common and conspicuous. Most are typically bee-like in appearance, being of medium size and having a compact body. Colour varies considerably among the various species.

The large and more noticeable carpenter bees belong to the genus *Xylocopa*. These insects chew long tunnels into dry wood or use the hollow stems or branches of plants as nesting sites. The bees partition off the tunnels into a series of cells using wood scrapings mixed with saliva. Before closing each cell, the female stocks it with a pasty mixture of pollen and nectar, on top of which she lays a single, large egg. The larva feeds on its supply of food until it is fully grown, then pupates within the cell, and finally, as an adult, chews its way to the outside world. Female carpenter bees can sting, but despite their size their sting is not as painful as that of the ordinary honeybee.

A range of smaller bees of genera such as *Ceratina* and *Allodape* also belong to this family. *Ceratina* spp. have essentially the same nesting habits as *Xylocopa*, but on a smaller scale as befits their size. The allodapine bees also use hollow twigs and stems for nesting sites, but do not make partitions to form cells or compartments. The female lays a few eggs in the tunnel and waits until the larvae emerge. She then provides food, a mixture of pollen and nectar, for her offspring. In between foraging trips she also tends to the other needs of the developing young, such as ensuring a clean nest by removing waste.

During the pupal stage the female remains on guard to fend off predators and parasites. After they have emerged as adults, some of the offspring fly off to start new nests elsewhere while some remain and, together with the now ragged-looking 'queen', start a new cycle of eggs. This can be regarded as a primitive stage in the development of true social life and energy-conserving division of labour, as exhibited by honeybees.

Xylocopids are widespread and common in all the warmer, well-vegetated areas of Africa. Several species occur over the whole length of the KNP.

PLATE 42

Family Apidae (Honeybees and other social bees)
Heuningbye

The members of this family are generally medium-sized but include several small and a few large species. The importance of the Apidae lies in the abundance and effectiveness of many species as pollinators of flowers and, to a lesser degree, as honey producers. All species are social, living in communities where division of labour has resulted in certain segments of the colony having specific tasks to perform.

Of the many species the best known is the honeybee, *Apis mellifera*, which has several subspecies, all differing slightly in habits. Much research has been done on this species, and volumes of fascinating information are available on the biology of this ancient associate of man. A typical colony is made up of one (rarely more than one) queen, many workers (there may be more than 50 000 in a flourishing colony) and drones (male bees in variable numbers, but never very many). The sole function of the drones is to mate with the queen. The queen lays the eggs, and produces pheromones which to a large extent regulate and ensure the smooth functioning of the community. She may lay more than 1 500 eggs a day (in summer). Most of these she fertilizes with the sperm which she stores in her spermatheca, but she deposits some eggs unfertilized; these eggs produce drones, while those which are fertilized develop into workers or queens, depending on the makeup of the food the growing larvae receive. After hatching the larvae are fed a secretion, often referred to as 'royal jelly', produced by glands in the head of mature adult workers. After a few days their diet is changed to honey and pollen, and these larvae then become workers; a few larvae, however, are kept on the diet of royal jelly and become queens. As the young workers grow older their wax-producing glands become functional and they turn to building or repairing the cells making up the comb. They also receive pollen and nectar collected by foraging workers, convert the nectar into honey and store it in the cells. Young worker adults are responsible for feeding the larvae developing in the cells of the comb. The final stage in the worker's life is spent searching for nectar, pollen, water and resin outside the hive.

Communication and direction-finding is highly developed in the honeybee. Workers flying in search of flowers orientate themselves by means of the polarized light coming from the sun, continually compensating for the changing position of the sun as the day progresses, so that when returning to the hive they can do so in a straight line no matter how twisted a course they flew on the outward path. Equally fascinating is the ability of every worker to communicate to its fellows in the hive, the direction and distance of a good supply of food. This they do by a series of body movements or dances which can be varied in intensity to indicate distance and are done in a particular plane of movement to show direction.

Honeybees and bees of the genus *Trigona* (also known as mopane bees) are widespread and common over the tropical parts of Africa. Honeybees are common in the KNP, mostly constructing nests in rock crevices or in hollow trees, and workers are frequently seen at flowers in all the camps or converging around soft-drink bottles and at the edges of pools of water on hot days. *Trigona* spp., all of which lack a sting, having vestigial, non-functioning stinging apparatus, are often seen in the northern parts of the KNP. They measure approximately 3 mm in length, are black in colour, and live in communities which usually nest in mopane trees (*Colophospermum mopane*). Several species of *Trigona* appear to occur in the KNP, one nesting in the hollows of *Androstachys johnsonii* trees in the northern areas, while some nest in termite mounds. The nectar of the flowers of *Androstachys johnsonii* contains substances which induce hallucinations in humans, having an effect somewhat like the drug mescaline, and the honey of the associated bees is avoided by some rural people who claim it makes them 'mad'!

PLATE 43

Family Formicidae (Ants)
Miere

A large number of species is included in this very specialized and highly developed family. All ants have elbowed antennae, but their most prominent feature is the constriction of the second (often also the third) abdominal segment into a narrow, scale-like or bead-like node to form a conspicuous petiole immediately behind the last pair of legs. Most species of the Formicidae are brown or black in colour and all live in communities generally composed of queens, workers, males and, sometimes, soldiers.

In almost all species the males are winged, have a small head with weak mandibles, and long, slender antennae. The queen is usually large compared to other members of her species, and begins life as a winged insect but loses her wings soon after mating. Workers are the most numerous members of the colony, are wingless, and have a small thorax and small eyes. The workers of many species are capable of laying eggs but these rarely hatch and are instead fed to the larvae as food. In some species a proportion of the workers have a large head and powerful mandibles; these individuals do the work of soldiers, being entrusted with the defence of the colony.

Ants are often compared with termites (page 70) despite their belonging to totally different orders and differing in many structural and biological aspects. Both groups are social insects, are divided into various castes, and have certain similar habits. These resemblances stem from both groups having independently adopted a communal way of life, with caste-formation and division of labour following almost inevitably to reduce energy expenditure. Much as in the case of termites, colony founding in ants – if both sexes are winged – starts with a mass emergence of males and females from existing colonies. It appears that this emergence is triggered by climatic factors, as all the nests of a particular species in an area emit swarms of these reproductive individuals at the same time. Following this 'nuptial flight', copulation takes place in the air, and when the fertilized females land they shed their wings, burrow into the soil and so start new nests. Each female digs a small chamber and lays a few eggs, following which (with a few exceptions) she seals herself in the chamber until the eggs hatch, living on her fat reserves and degenerating flight muscles. The larvae are fed on secretions from the salivary glands of the queen. and when they are fully grown they exit from the chamber to forage for food outside. All these initial offspring are workers which take over the foraging and cleaning duties, enlarge the nest, and allow the queen to commence with her main function of egg-laying.

There are eleven subfamilies of ants, each of which has reached a particular level of evolutionary advancement and may be recognized by its distinctive habits. The most primitive are the Ponerinae and the Myrmeciinae. Among the Ponerine ants are the Matabele ants, *Megaponera foetens*, which are often seen moving in columns usually one or two metres long but in exceptional cases exceeding 20 metres, the ants marching between two and 10 ants abreast. Each individual ant measures between 10 mm and 20 mm depending on the caste. The colony has an underground nest and from this base makes periodic raids, either at night or on overcast days, on places where termites are concentrated. Termites appear to be their only prey, and after a successful raid each ant can be seen with up to five termites gripped in its mandibles. These termites will be ripped apart and fed to the queen, the males, the larvae and other members in the nest. If Matabele ants are disturbed in any way they immediately break rank and mill around in search of the cause of the disturbance, all the while venting a daunting, high-pitched squeaking noise made by rubbing certain body segments against each other, and will readily administer a very painful sting if given the opportunity. Wandering bands of Matabele ants are usually seen in the far north of the KNP, most abundantly at Pafuri where they are of considerable importance in limiting the numbers of termites so numerous in that area.

The Dorylinae are slightly more advanced and include the army ants, also known as driver ants or legionary ants. These are carnivorous species which prey on other insects and small animals, but will readily scavenge on dead animals. They do not have permanent nests, but lead a nomadic life, wandering around in columns at night or on cool days searching for prey, making temporary nests in the soil until their hunting activities have depleted most of the prey in the area, then moving on to a new location. The colonies may be enormous, numbering tens of thousands of reddish-brown ants, each measuring between 2 mm and 8 mm in length, although Skaife et al (1979) claim that some colonies may contain as many as 20 million individuals. The queens of *Dorylus helvolus* measure up to 50 mm in length, and have an enormously distended abdomen filled with the reproductive organs, capable of laying four million eggs a month. Male driver ants of this species are often seen flying clumsily around lights on summer evenings in all the camps in the KNP. They measure between 25 mm and 30 mm in length and may be recognized by their long, sausage-like, reddish-brown abdomen. Although they twist the abdomen menacingly when held, they have no sting and rely completely on bluff. *Dorylus helvolus* is widely distributed throughout Africa; whole colonies have been seen by the author at Pafuri and Skukuza, swarming in vast numbers around carcasses in search of insects congregating there.

With the exception of some species in the Myrmicinae, all the members of the following subfamilies have adopted a vegetarian diet in the form of either seeds, fruit and other solids, or liquids such as nectar and honeydew.

The Myrmicinae are an extraordinarily abundant and widespread group, including a large number of species. They have a double-jointed petiole. Many measure approximately 5 mm or considerably less in length. Included in this subfamily are the common blackish ants which can be seen crawling around the walls in the various camps. The 'cocktail ants' (named for their conspicuous habit of raising the abdomen when they are alarmed) of the genus *Crematogaster* belong to this subfamily. They construct nests in hollow twigs, in holes in stems, in hollow acacia thorns or under bark, or build elaborate carton nests made of finely chewed plant fibres mixed with saliva and built around twigs. In the KNP these carton nests have been noticed only south of the Olifants River, and even in this area are not common. If such a nest is tapped, hundreds of ants stream agitatedly out of the multitudinous tunnels of the nest, abdomens raised threateningly at the intruder. Although they do possess a sting they rarely use it, relying instead on the repellant secretion which is emitted at the hind end of the abdomen. Ants of the genus *Crematogaster* are widespread throughout the KNP, normally seen on vegetation in search of food such as the sweet secretions given off by many sap-sucking Hemiptera such as aphids, coccids and membracids.

The subfamily Formicinae is the most highly advanced and specialized group in terms of structure and habits. Several species have refined the art of food collecting to a remarkable degree. Some species which feed on honeydew and other insect secretions have taken to tending certain aphids and coccids in much the same way as man tends cattle. The ants protect these insects from predators and parasites, sometimes constructing shelters around them, and in winter carry their eggs back to the nest to ensure a following generation. Other species, known as 'repletes' (not found in the KNP) have adopted a strikingly unusual way of storing food: some of the workers are fed with nectar and honeydew until their abdomens swell enormously, resembling – and for all practical purposes being – no more than storage pots from which the colony obtains its food in times of scarcity. A very common genus of formicine ants is *Camponotus*, some of which are called 'sugar ants', one of the largest and most widespread genera in Africa. Most are active at night and, depending on the species, construct their nests under stones, logs or bark, or within dead trees, while some species build carton nests or silk nests from silk produced by the larvae.

Africa has a very wide range of ant species, a large number of which occur abundantly and are widespread in the KNP.

Glossary

Terms used in this book as they pertain to insects. (The Afrikaans term appears (*in italics*) directly after the English term.)

Abdomen *(Abdomen)* Hindmost of the three major body divisions, after the head and thorax; contains the digestive and reproductive organs.

Adecticous *(Adekties)* Not having functional mandibles. Term is usually applied to the pupal stage of an insect where it may or may not require mouthparts to assist it in escaping from the puparium or cocoon. *See* Decticous.

Anal lobe *(Anaallob)* One of four main regions of the wing, lying between the anal and jugal fold and being roughly triangular in shape.

Antennae *(Antennae)* Paired 'feelers' attached to the head, having a sensory function. Chemoreceptors, which enable the insect to 'smell', are present on the antennae.

Apical *(Apikaal)* Terminal; at the tip or end of the limb or structure. *See* Distal, Proximal.

Aposematic coloration *(Aposematiese kleure)* The phenomenon of insects using bright body colours to warn predators of distasteful or poisonous tissues and secretions.

Appendix dorsalis *(Appendix dorsalis)* A median filament of uncertain purpose projecting posteriorly from the abdomen of Archaeognatha, Thysanura and some Ephemeroptera.

Apterous *(Vlerkloos)* Lacking wings.

Arolium *(Arolium)* Small median lobe or pad between the claws at the tip of some insects' legs, forming part of the pretarsus.

Arthropod *(Arthropod)* Invertebrate animal having jointed legs; includes (among others) the Crustacea, Myriapoda, Arachnida and Insecta.

Atrophy *(Atrofeer)* Reduction in size and function of a muscle, organ or other body part, generally through lack of use.

Batesian mimicry *(Batesiaanse mimikrie)* When a harmless species mimics the body shape and bold warning coloration of a distasteful species so as to avoid predators. *See* Müllerian mimicry.

Brachypterous *(Kortvlerkig)* Having short wings or wings much reduced in size.

Carina *(Kiel)* Keel-shaped part or ridge.

Cercus *(Serkus)* One of a pair of jointed appendages at the distal end of the abdomen of many insects, having a sensory function. (*Pl.* Cerci.)

Chitin *(Chitien)* A nitrogenous carbohydrate derivative making up much of the cuticle of insects and other arthropods.

Clypeus *(Klipeus)* A sclerite between the frons and labrum on the head; a small, hardened plate more or less overlying the insect's mouth.

Costal cell *(Kostaalsel)* Part of the insect wing.

Coxa *(Koksa)* Basal segment or proximal joint of insect leg; that segment which joins the leg to the thorax. (*Pl.* Coxae.)

Crawlers *(Kruipers)* First-stage larvae of scale insects.

Crepuscular *(Krepuskulêr)* Dusk-flying.

Cursorial *(Kursoriaal)* Adapted for running. *See also* Fossorial, Gressorial, Saltatorial, Raptorial.

Cuticle *(Kutikula)* The external covering of the insect which continues into the fore- and hind gut and also lines the ducts of the dermal glands and the entire tracheal system. It is made up of a thin, nonchitinous outer epicuticle, and a thicker, chitinous inner procuticle.

Decticous *(Dekties)* Having functional mouthparts which can be used to open the puparium or cocoon. *See* Adecticous.

Dichoptic *(Dichopties)* Compound eyes not touching or joined medially on the head. *See* Holoptic.

Distal *(Distaal)* That part furthest from the base. *See* Apical, Proximal.

Diurnal *(Daglewend)* Active in daylight.

Ecdysis *(Ekdiese)* The process of moulting or shedding the cuticle. (*Adj.* Ecdysial.)

Ectoparasite *(Ektoparasiet)* A parasite occurring externally on its host (eg. a tick or louse). *See* Endoparasite.

Elytron *(Elytron)* Hardened sheath-like forewing present in the Coleoptera. (*Pl.* Elytra.) *See* Hemelytron, Tegmen.

Endoparasite *(Endoparasiet)* A parasite occurring inside its host (eg. a tapeworm). *See* Ectoparasite.

Entognathous *(Entognaat)* Having the mouthparts enclosed laterally by pleural folds which grow down from the sides of the head to fuse with the labium. *See* Exognathous.

Eruciform *(Erusiform)* Resembling a caterpillar.

Exarate *(Eksaraat)* A pupa having wings and legs loose from the body. *See* Obtect.

Exognathous *(Eksognaat)* Having the mouthparts exposed, situated outside the oral cavity; is a characteristic used to distinguish true insects from other groups such as the Collembola, Protura and Diplura. *See* Entognathous.

Femur *(Femur)* A segment of the leg between the trochanter and tibia. *(Pl.* Femora.)
Filiform *(Filiform)* Thread-like. *See* Moniliform.
Flabellate *(Flabelaat)* Fan-shaped.
Fontanelle *(Fontanel)* A pale, depressed spot found on the front of the head of some termites.
Fossorial *(Fossoriaal)* Adapted for digging. *See also* Cursorial, Gressorial, Raptorial, Saltatorial.
Frons *(Frons)* That part of the upper or anterior surface of the head lying between the eyes; corresponds more or less with the forehead.
Galea *(Galea)* A component of the external mouthparts; part of the maxilla.
Genitalia *(Genitalia)* External reproductive organs.
Gressorial *(Gressoriaal)* Adapted for walking. *See also* Cursorial, Fossorial, Raptorial, Saltatorial.
Haemolymph *(Haemolimf)* The invertebrate equivalent of vertebrate blood. The composition differs, however, and in most cases haemolymph is not used in respiration.
Haltere *(Halteer)* Club-like balancing organ; the modified hind wings of the Diptera.
Haustellum *(Haustellum)* Mouthparts adapted to form an elongate proboscis used for sucking.
Hemelytron *(Hemelytron)* The forewing of some Hemiptera which has one half hardened to protect the underlying, membranous hind wing. *See* Elytron, Tegmen.
Hemimetabolous *(Hemimetabolies)* Insect with a life cycle of incomplete or partial metamorphosis. No dramatic change in body form occurs between the immature and adult life stages, the immatures resembling the adults in body shape. *See* Holometabolous.
Hexapoda *(Hexapoda)* Animals having six legs.
Holometabolous *(Holometabolies)* Insect with a life cycle involving complete metamorphosis. There is a dramatic change in body form between the immature and adult life stages, the adults appearing completely different to the immatures, with the change generally occurring in the pupal stage. *See* Hemimetabolous.
Holoptic *(Holopties)* Compound eyes joined or meeting in the middle of the head. *See* Dichoptic.
Hypognathous *(Hipognaat)* Mouthparts directed ventrally on the head. *See* Opisthognathous, Prognathous.
Imago *(Imago)* The adult stage in insect metamorphosis.
Instars *(Instars)* Larval or nymphal stages.
Labium *(Labium)* One of the mouthpart components ventral to the mouth opening; comparable to a lower lip.

Labrum *(Labrum)* One of the mouthpart components, generally a moveable plate; comparable to an upper lip.
Lamellate *(Lamellaat)* Plate-like or scale-like.
Larva *(Arf)* Immature or juvenile stage of holometabolous insect after it has emerged from the egg.
Mandible *(Mandible)* The biting or chewing component of insect mouthparts; may be highly modified in insects which do not bite or chew.
Maxilla *(Maksilla)* Part of the mouthparts, posterior to the mandibles, that assist the mandibles as accessory jaws to hold and chew food.
Mesonotum *(Mesonotum)* The dorsal sclerotized plate covering the second thoracic segment. *See* Notum.
Mesothorax *(Mesotoraks)* Middle thoracic segment.
Metamorphosis *(Metamorfose)* Change of form; generally refers to the dramatic changes from larvae to pupae to adults. *See also* Hemimetabolous and Holometabolous.
Metanotum *(Metanotum)* The dorsal sclerotized plate covering the third (hindmost) thoracic segment. *See* Notum.
Metathorax *(Metatoraks)* Last thoracic segment.
Microtrichia *(Mikrotrichia)* Small hairs lacking basal articulation, found on insect wings.
Moniliform *(Monoliform)* Generally referring to antennae which resemble a string of beads; antennae strongly constricted at regular intervals. *See* Filiform.
Müllerian mimicry *(Mülleriaanse mimikrie)* When two distasteful species have the same bold coloration and patterns to reinforce a warning and repellant effect on predators. *See* Batesian mimicry.
Nocturnal *(Naglewend)* Active at night.
Notum *(Notum)* The dorsal sclerotized plate covering the pronotum, mesonotum, and metanotum. Each notum is generally divided into an anterior prescutum, a scutum, and posterior scutellum.
Nymph *(Nimf)* The immature stage of hemimetabolous insects; the stage between egg and adult. *(Pl.* Nymphae.)
Obtect *(Obtektaat)* A pupa having the wings and legs held closely against the body. *See* Exarate.
Ocellus *(Osellus)* Simple eye made up of a single ommatidium or a small number of ommatidia, as opposed to the large compound eyes which consist of numerous grouped ommatidia. Most insects have two compound eyes, and one or more simple eyes arranged dorsally between the compound eyes. *(Pl.* Ocelli.) *See* Stemma.

Ommatidium *(Ommatidium)* One of the component elements making up the insect eye; a single unitary eye which together with many others collectively make up the compound eye of an insect. (*Pl.* Ommatidia.)

Ootheca *(Ooteka)* Egg pod.

Opisthognathous *(Opistognaat)* Head arranged in such a manner that the mouthparts are held backwards, pointing posteriorly. *See* Hypnogathous, Prognathous.

Osmeterium *(Osmeterium)* A forked protrusible organ which emits a repellent odour, occurring on the first thoracic segment of some butterfly larvae. (*Pl.* Osmeteria.)

Oviparous *(Ovipaar)* Egg-laying.

Ovipositor *(Eierboor)* Egg-laying apparatus found at posterior end of female insect's body.

Palp *(Palpus)* The segmented sensory appendages on the maxillary or labial mouthparts.

Parasite *(Parasiet)* An organism temporarily or permanently occurring on or within another organism. Although it deprives the host of some nutrients for its sustenance, it does not cause the death of the host.

Parasitoid *(Parasitoïed)* A form of parasitism which borders on predation, in which the insect feeds on the host tissue with little initial harm but which eventually leads to the death of the host.

Parthenogenesis *(Partenogenese)* Females producing viable offspring without having been fertilized by males.

Petiole *(Petiool)* A slender stalk connecting the thorax and abdomen of some insects, mostly in wasps. Also called a pedicel.

Pheromones *(Feromoon)* Chemical substances released from the body of the insect, used to communicate with other insects and influence their behaviour or even metabolism. They may serve to bring sexes together for reproduction, elicit an alarm response, promote co-operative behaviour among social insects and a range of other responses.

Phytophagous *(Fitofaag)* Plant-feeding.

Pleuron *(Pleuron)* The sclerotized plate on the side of each segment of the insect body. (*Pl.* Pleura.) *See* Sternite, Tergite.

Post-clypeus *(Postklipeus)* The posterior part of the clypeus of an insect.

Pretarsus *(Pretarsus)* The end part of the insect foot, generally made up of a ventral plate, a pair of claws and the arolium.

Proboscis *(Proboskis)* Mouthparts modified into an elongate, tube-like or trunk-like structure.

Prognathous *(Prognaat)* Mouth opening situated anteriorly on the head. *See* Hypognathous, Opisthognathous.

Proleg *(Propoot)* False legs found in addition to the three pairs of true legs in the immature stages of many insects. Also called a pseudopod.

Pronotum *(Pronotum)* The upper surface of the first thoracic segment behind the head of insects.

Propleuron *(Propleuron)* The cuticle on the sides of the prothorax.

Prothorax *(Protoraks)* The first or most anterior segment of the thorax; the first segment behind the head.

Proximal *(Proksimaal)* Nearest the body or point of attachment. *See* Apical, Distal.

Pseudohaltere *(Pseudohalteer)* False halteres found in males of the Coccoidea, where the hind wings are reduced to club-like structures resembling those of the Diptera.

Pseudopod *(Pseudopoot) See* Proleg.

Pterostigma *(Pterostigma)* Opaque wing spot; generally refers to a dark spot or marking on the anterior or leading edge of the wing.

Pulvillus *(Pulvillus)* The pad beneath a segment or segments of the tarsus of the insect leg, or between the claws of the tarsus. (*Pl.* Pulvilli.)

Pupa *(Papie)* One of the life-stages of holometabolous insects; the 'resting stage' between the larval and adult stages, during which major metamorphic changes occur, larval organs and structures degenerate and adult tissues grow from groups of cells previously present but dormant.

Puparium *(Puparium)* The hardened skin of the final-instar larva, which forms a protective case within which the pupa lies. (*Pl.* Puparia.)

Raptorial *(Gryppote)* Adapted to seize and hold prey. *See also* Cursorial, Fossorial, Gressorial, Saltatorial.

Remigium *(Remigium)* The major region of the wing responsible for propulsion and maintenance of flight. The other wing regions are the anal area, the axillary area and, in some insects, the jugal area (also called neala).

Rostrum *(Rostrum)* Mouthparts modified into a beak-like prolongation.

Saltatorial *(Saltatoriaal)* Adapted for jumping. *See also* Cursorial, Fossorial, Gressorial, Raptorial.

Sapwood *(Spinthout)* The pale, soft wood lying below the bark of trees.

Sclerites *(Skleriete)* Plates of the hardened cuticle making up the outer covering of the insect.

Sclerotized *(Gesklerotiseerd)* The soft hydrated protein and chitin of the new insect cuticle converted into the typical hardened and tough outer covering normally protecting the insect.

Setaceous *(Behaard)* Having bristles or hair-like protrusions on the body.

Setae *(Setae)* Bristles or hair-like structures.

Spinneret *(Spinneret)* Organ perforated by tubes and connected with glands secreting liquid silk, used to spin silk.

Spiracles *(Spirakel)* Openings of the tracheae on the side of the body through which oxygen moves in and carbon dioxide out; used for purposes of respiration.

Spurs *(Spore)* Spiny projection articulating from the base, generally found on the tibia.

Stemma *(Stemma)* Simple eye. (*Pl.* Stemmata.) *See* Ocellus.

Sternite *(Sterniet)* A component plate making up the sternum.

Sternum *(Sternum)* The covering of the ventral surface of each body segment; each sternum may be made up of a single sternite or multiple sternites. (*Pl.* Sterna.) *See* Pleuron, Tergum.

Stylets *(Stillet)* Pointed, bristle-like appendages.

Styliform *(Styliform)* Tubular.

Sutures *(Sutare)* Line of junction between two sclerotized plates of the cuticle.

Tarsus *(Tarsus)* Part of the insect leg making up the major part of the 'foot', found between the tibia and pretarsus, and itself made up of several tiny segments called tarsomeres. (*Pl.* Tarsae.)

Tegmen *(Tegmen)* Generally referring to the modified forewings of the blattoid-orthopteroid orders, where the normal soft membranous forewing is hardened as a protective sheath to cover the large, soft hind wings. Also refers to a component part of the male copulatory organ. (*Pl.* Tegmina.) *See* Elytron, Hemelytron.

Tergite *(Tergiet)* A hardened plate making up all or part of the tergum. *See* Sternite.

Tergum *(Tergum)* The dorsal covering of each thoracic and abdominal segment. *See* Pleuron, Sternum.

Thorax *(Toraks)* Second of the three main body divisions in insects, between the head and the abdomen. The true legs, and wings in those insects which are winged, are attached to the thorax. The thorax is subdivided into three segments, the most anterior being the prothorax, followed by the mesothorax and the metathorax.

Tibia *(Tibia)* One of the largest segments making up the insect leg, found between the femur and the tarsus. (*Pl.* Tibiae.)

Tracheae *(Trachea)* A set of fine tubes running throughout the insect and ramifying to almost every cell of the body, connected to the external environment by spiracles, and through which oxygen is conveyed to the tissues and carbon dioxide removed.

Triungulin *(Triungulin)* The first-instar larva of some beetle families. Free-living, active insects adapted to find hosts they can parasitize.

Trochanter *(Trochanter)* A small segment making up part of the insect leg, found between the coxa and the femur.

Veins *(Are)* Hardened tubes which form the strengthening and supporting framework of the insect wing. There is a basic pattern of main veins, their branching and interconnections varying according to group and even species.

Venation *(Beaaming)* The particular pattern which the veins on the wing assume; the venation of the wing is an important diagnostic character, used to separate different groups of insects.

Vestigial *(Vestigiaal)* Imperfectly developed or reduced in comparison with the size of the organ or structure in its original, ancestral state.

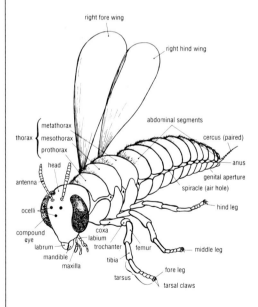

Stylized representation of the external structure of a typical insect.

Bibliography

ASIBEY, E.O.A. 1977. 'The Blackfly Dilemma'. *Environmental Conservation* 4(4): 291-295.

BENGIS, R.G. 1982. 'An Examination for Nagana and Tsetse Flies in the KNP'. *Annual Report, Kruger National Park, 1981-1982.* p.78.

BRAACK, H.H., DAVIDSON, I.H., LEDGER, J.A. and LEWIS, D.J. 1981. 'Records of sandflies (Diptera: Psychodidae. Phlebotominae) feeding on amphibia, with a new record from the Kruger National Park'. *Koedoe* 24:187-188.

BRAACK, L.E.O. 1979. 'Insects and Carcasses'. *Custos* 8(6):6-15.

BRAACK, L.E.O. 1979. 'For the Gourmet'. *Custos* 8(11):25-26.

BRAACK, L.E.O. 1981. 'Visitation patterns of principal species of the insect complex at carcasses in the Kruger National Park'. *Koedoe* 24:33-49.

BRAACK, L.E.O. 1983. 'Most successful of all living organisms. Insects. Part 1'. *Custos* 11(12):6-9.

BRAACK, L.E.O. 1983. 'Darting dragonflies in fact predators'. *Custos* 12(2):23-26.

BRAACK, L.E.O. 1983. 'Eating enough for 60 000 cattle'. *Custos* 12(3):6-7.

BRAACK, L.E.O. 1983. 'The cricket's chirr is a mating call'. *Custos* 12(4):20-24.

BRAACK, L.E.O. 1983. 'Don't call any bug a "bug"'. *Custos* 12(7):16-17.

BRAACK, L.E.O. 1984. 'Ladybird predators: the gardener's friends'. *Custos* 12(10):19-22.

BRAACK, L.E.O. 1984. 'These two types of flies the top disease carriers'. *Custos* 13(1):16-19.

BRAACK, L.E.O. 1984. 'Of bats, bugs, caves and superstition'. *Custos* 13(3):15-16.

BRAACK, L.E.O. 1984. *The insects associated with exposed carcasses in the northern Kruger National Park. A study of populations and communities.* Unpublished Ph.D. thesis, University of Natal, Pietermaritzburg. 260 pp.

BRAACK, L.E.O. 1984. 'Mortality of blow-fly larvae (Diptera: Calliphoridae) in the digestive tract of vultures'. *Koedoe* 27:5-8.

BRAACK, L.E.O. 1984. '"Epidermal streaming" and associated phenomena displayed by larvae of *Chrysomya marginalis* (Wd) (Diptera: Calliphoridae) at carcasses'. *Koedoe* 27:9-12.

BRAACK, L.E.O. 1984. 'A note on the presence of the elephant louse *Haematomyzus elephantis* Piaget (Mallophaga: Rhynchophthirina) in the Kruger National Park'. *Koedoe* 27:139-140.

BRAACK, L.E.O. 1985. Disease transmission potential of blow-flies (Diptera: Calliphoridae) in the Kruger National Park. *Proceedings of the 5th Congress of the Entomological Society of Southern Africa.* p.4-5.

BRAACK, L.E.O. 1986. 'Curious quirks in a miniature world'. *Custos* 14(12):12-15.

BRAACK, L.E.O. 1986. 'The Bats and bugs of Lanner Gorge'. *Custos* 15(5):30-32.

BRAACK, L.E.O. 1986. 'Arthropods associated with carcasses in the northern Kruger National Park'. *South African Journal for Wildlife Research* 16(3):91-98.

BRAACK, L.E.O. and RETIEF, P.F. 1986. 'Dispersal, density and habitat preference of the blow-flies *Chrysomyia albiceps* (Wd) and *Chrysomya marginalis* (Wd) (Diptera: Calliphoridae)'. *Onderstepoort Journal of Veterinary Research* 53:13-18.

BRAACK, L.E.O. and EMERSON, K.C. 1986. 'A louse phoretic on a haematophagous muscid fly'. *Journal of the Entomological Society of Southern Africa* 49(1):161-162.

BRAACK, L.E.O. 1987. '*Musca marginalis* Wiedemann, 1830, (currently *Chrysomya marginalis*: Insecta, Diptera): proposed conservation of the specific name'. *Bulletin of Zoological Nomenclature* 44(1):13-14.

BRAACK, L.E.O. 1987. 'Arthropod utilisation of a bat-frequented cave in the Kruger National Park'. *Proceedings of the 6th Congress of the Entomological Society of Southern Africa.* p.8.

BRAACK, L.E.O. 1987. 'Community dynamics of carrion-attendant arthropods in tropical African woodland'. *Oecologia* (Berlin) 72:402-409.

BRAACK, L.E.O. and DE VOS, V. 1987. 'Seasonal abundance of carrion frequenting blow-flies (Diptera: Calliphoridae) in the Kruger National Park'. *Onderstepoort Journal of Veterinary Research* 54:591-597.

BRAACK, L.E.O. 1988. *The Kruger National Park: A Visitor's Guide.* (2nd Edition.) Struik Publishers, Cape Town.

BRAACK, L.E.O. 1989. 'National Parks Board's locust control campaign'. *Custos* 18(8): 22-26.

BRAACK, L.E.O. 1989. 'Arthropod inhabitants of a tropical "island" environment provisioned by bats'. *Biological Conservation* 68: 77-86.

BRAACK, L.E.O. 1990. 'War refugees and malaria epidemiology: the South Africa/Mozambique situation'. *Proceedings of the VIIth International Congress of Parasitology, Paris, France.* p.665.

BRAACK, L.E.O. and DE VOS, V. 1990. 'Feeding habits and flight range of blow-flies in relation to anthrax transmission in the Kruger National Park, South Africa'. *Onderstepoort Journal of Veterinary Research.* 57: 161-162.

CILLIERS, C. and REID, P. 1987. 'Kewerwyfies oorwin in Krugerwildtuin'. *Custos* 15(11):20-21.

DALY, H.V., DOYEN, J.T. and EHRLICH, P.R. 1978. *Introduction to insect biology and diversity.* McGraw-Hill, New York. + 564 pp.

DE MEILLON, B., and WIRTH, W.W. 1981. 'Sub-Saharan Ceratopogonidae (Diptera) VII. The biting midges of the Kruger National Park, South Africa, exclusive of the genus *Culicoides*'. *Annals of the Natal Museum* 24(2):563-601.

GALIL, J. 1977. 'Fig Biology'. *Endeavour* 1(2):52-56.

GEAR, J.H.S., HANSFORD, C.F., and PITCHFORD, R.J. 1988. *Malaria in Southern Africa.* Department of National Health and Population Development, Pretoria.

GREENBERG, B. 1971. *Flies and Disease. Vol. 1 and 2.* Princeton University, Princeton. 856 + 447 pp.

HOLM, E. and DE MEILLON, E. 1986. *Insects.* Struik Pocket Guide Series, Struik Publishers, Cape Town.

HORAK, I.G. Numerous scientific articles on parasites of wildlife in the KNP and published mainly since 1970 in *The Onderstepoort Journal of Veterinary Research.* All these papers are bound as a single volume under the title of the next reference.

HORAK, I.G. 1989. *Studies on helminth and arthropod parasites of some domestic and wild animals in southern Africa.* D.VSc. thesis. University of Pretoria, Pretoria.

KLOPPERS, J. and VAN SON, G. 1978. *Butterflies of the Kruger National Park.* Board of Curators for National Parks of the Republic of South Africa, Pretoria.

KOLBE, F.F. 1974. 'The status of the tsetse flies in relation to game conservation and utilisation'. *Journal of the South African Wildlife Management Association* 4(1):43-49.

LEDGER, J.A. 1980. *The arthropod parasites of vertebrates in Africa south of the Sahara: Phthiraptera (Insecta).* The South African Institute for Medical Research, Johannesburg. 327 pp.

McALPINE, J.F. 1978. 'A new species of *Piophila* from South Africa (Diptera: Piophilidae)'. *Annals of the Natal Museum* 23(2):455-459.

McINTOSH, B.M. 1975. Mosquitos as vectors of viruses in southern Africa. *Department of Agricultural Technical Services, South Africa. Entomology Memoir.* No 43. 19 pp.

MEISWINKEL, R., BRAACK, L.E.O. and MEISWINKEL, P. 1987. 'Afrotropical *Culicoides*: Further research on the genus in the Kruger National Park with special emphasis on the larval habitats of the subgenus *Avaritia* in large mammal dung'. *National Parks Board of South Africa Research Report 1986/1987.* p.100-102.

NEWLANDS, G. 1989. *Malaria and mosquitos in Southern Africa.* Unibook Publishers, South Africa.

ODHIAMBO, T.R. 1977. Entomology and the problems of the tropical world. *Proceedings of the XV International Congress of Entomology.*

RAUTENBACH, I.L., KEMP, A.C. and SCHOLTZ, C.H. 1988. 'Fluctuations in the availability of arthoropods correlated with microchiropteran and avian predator activities'. *Koedoe* 31: 77-90.

RICHARDS, O.W. and DAVIES, R.G. 1977. *Imm's general textbook of entomology. Volume 2.* Chapman and Hall, London. p.421-1354.

SCHOLTZ, C.H. and HOLM, E. (eds). 1985. *Insects of Southern Africa.* Butterworths, Durban.

SKAIFE, S.H., LEDGER, J. and BANNISTER, A. 1979. *African insect life.* C. Struik, Cape Town. 279 pp.

SOUTHWOOD, T.R.E. 1977. Entomology and mankind. *Proceedings of the XV International Congress of Entomology.*

STEVENSON-HAMILTON, J. 1974. *South African Eden.* William Collins, London.

STUCKENBERG, B.R. and STUCKENBERG, P.J. 1974. Report on a collecting trip to the Kruger National Park. *Unpublished report, Skukuza, Kruger National Park.* 9 pp.

THOMSON, G.R., DOUBE, B.M., BRAACK, L.E.O., GAINARU, H.D. and BENGIS, R.G. 1988. 'Failure Of *Haematobia thirouxi potans* (Bezzi) to transmit foot-and-mouth disease virus mechanically between viraemic and susceptible cattle'. *Onderstepoort Journal of Veterinary Research* 55: 121-122.

WEAVING, A. 1977. *Insects: a review of insect life in Rhodesia.* Regal, Salisbury. 179 pp.

WEAVING, A. 1982. 'Dung beetles'. *The Naturalist* 26(1):p.2-6.

ZIEGLER, P. 1969. *The Black Death.* Collins, London. 319 pp.

ZUMPT, F. (ed.). *The arthropod parasites of vertebrates in Africa south of the Sahara (Ethiopian Region). Volume III (Insects excluding Phthiraptera).* The South African Institute for Medical Research, Johannesburg. 283 pp.

INDEX TO COMMON NAMES

Numbers in **bold** indicate the main entry, and numbers in *italics* refer to the colour plates.

alderflies **91**, *Pl. 12*
antlions 91, **92-93**, *Pl. 13*
ants 67, 70, 71, 84, 85, 97, 99, 115, 133, 135, **142-145**
  army 144
  cocktail 144
  driver 144, *Pl. 43*
  flying 70
  legionary 144
  Matabele 71, 143, *Pl. 43*
  sugar 145
  white 70
aphids 82, **85**, 92, 101, 115, 133, 135, 138, 144, *Pl. 8*

backswimmers **88-89**, 89, *Pl. 11*
bees 70, 114, 124, 135, 136, 138
  carder **140-141**
  carpenter 114, **141**, *Pl. 42*
  honey- 124, 135, 140, 141, **141-142**, *Pl. 42*
  leafcutter **140-141**, *Pl. 41*
  mopane 142
  social 141
beetles 11, 92, **95-108**, 139
  bark **106-107**, *Pl. 22*
  blister **102**, *Pl. 19*
  checkered **99-100**, *Pl. 17*
  Christmas **82-83**, *Pl. 7*
  click **100**, *Pl. 17*
  darkling **102-103**, *Pl. 19*
  diving **97**, 98, *Pl. 15*
  dung 103, **103-104**, *Pl. 19*
  engraver **106-107**, *Pl. 22*
  ground **96**, *Pl. 15*
  hide **101**, *Pl. 18*
  jewel **100-101**, *Pl. 18*
  ladybird **101**, *Pl. 18*
  leaf **106**, *Pl. 22*
  long-horn 99, **105-106**; *Pl. 21*
  lymexylid *Pl. 17*

  net-winged **99**, *Pl. 16*
  powderpost **101-102**
  rhinoceros 103, **104**, *Pl. 20*
  rove **98-99**, *Pl. 16*
  skin **101**, *Pl. 18*
  tiger **96**, *Pl. 15*
  tortoise 106, *Pl. 22*
  trogid *Pl. 20*
  whirligig **97-98**, *Pl. 15*
  woodboring 100
blues **132-133**, *Pl. 36*
booklice **79**, *Pl. 5*
borers, maize stalk 130,
  metallic wood- **100-101**, *Pl. 18*
  shot-hole **101-102**, *Pl. 19*
bristletails **65-66**
browns **135**, *Pl. 38*
bugs **82-90**
  assassin 82, 86, 87, **87**, *Pl. 10*
  bat **87-88**
  bed 82, **87-88**, *Pl. 10*
  burrowing **86**, 87, *Pl. 9*
  giant water **89-90**, *Pl. 12*
  plant **88**, *Pl. 10*
  seed **87**, *Pl. 9*
  shield **86**, 87, *Pl. 10*
  spittle **83**, *Pl. 7*
  squash **86-87**, *Pl. 9*
  stink 82, **86**
burnets **125-126**, *Pl. 31*
butterflies 109, 124, 137
  African monarch 134, *Pl. 38*
  brush-footed **133**, *Pl. 37*
  Christmas 131
  milkweed **134**, *Pl. 38*

caddisflies **123-124**, *Pl. 30*
chafer, flower *Pl. 20*
  leaf Pl. 20
charaxes **133-134**, *Pl. 37*
chiggers 95
chigoes 95
cicadas **82-83**, 84, *Pl. 7*

coccids 144
cockroaches **69-70**, *Pl. 2*
coppers **132-133**
coreids **86-87**, *Pl. 9*
cotton stainers **87**, *Pl. 10*
crickets 73, *Pl. 4*
  armoured ground 76, *Pl. 4*
  field **76-77**
  garden 75, 76
  ghost 76
  mole **77**, *Pl. 4*
  tree **76-77**, *Pl. 4*

damselflies **66-67**, *Pl. 1*
dragonflies **66-67**, 92, *Pl. 1*

earth measurers **127**, *Pl. 32*
earwigs **77-78**, *Pl. 5*

fireflies **99**, *Pl. 16*
fleas 8, **93-95**, *Pl. 14*
  bat *Pl. 14*
  cat 94
  dog 94
  human 94
  sand 95
  sticktight 94, *Pl. 14*
flies 85, 98, **109-123**, 138, 139
  bee **114-115**, 115, *Pl. 26*
  blow- 100, 101, 105, **118-120**, 120, 138, 140, *Pl. 28*
  bot **117-118**, *Pl. 27*
  crane 108, **109-110**, *Pl. 24*
  flesh **120**, *Pl. 28*
  horse **113-114**, 140, *Pl. 25*
  house 115, **115-116**, 140, *Pl. 27*
  hover **115**, *Pl. 26*
  lantern **83**, *Pl. 7*
  lesser house 116
  louse **121-122**, *Pl. 29*
  moth- **110**, *Pl. 24*
  putsi 119

152

redheaded **122**, *Pl. 30*
robber **114**, *Pl. 26*
sand- **110**
saw- **136**, *Pl. 39*
skipper **123-124**, *Pl. 30*
small 116
stalk-eyed **122**, *Pl. 30*
thick-headed **122**, *Pl. 30*
tsetse **120-121**, 121,
  139, *Pl. 29*
tumbu 119
warble **117-118**, *Pl. 27*
fulgorids **83**, *Pl. 7*

glow-worms 99, *Pl. 16*
gnats, gall **113**, *Pl. 25*
grasshoppers 73, **74-75**,
  114, *Pl. 3*
  long-horned 76

hairstreaks **132-133**
handmaidens **130**, *Pl. 34*
hanging flies **108-109**, *Pl. 23*
histerids **98**, *Pl. 16*

jiggers 95

katydids **76**, *Pl. 4*

lacewings 91, **91-92**, *Pl. 13*
leafhoppers **83-84**, *Pl. 7*
lice 8, **79-81**
  biting 80, *Pl. 6*
  body 81
  elephant *Pl. 6*
  head 81
  human 81
  plant- **85**, *Pl. 8*
  sucking 81, *Pl. 6*
locusts 73, **74-75**, *Pl. 3*
  brown 74, 75
  desert 74
  red 74, 75
loopers **127**, *Pl. 32*

mabungu grubs 105, *Pl. 21*
maggots, Congo floor 119
mantispids **92**, *Pl. 13*

mayflies **66**, *Pl. 1*
midges 11, **112**, *Pl. 25*
  biting **112**, *Pl. 24*
  gall **113**, *Pl. 25*
monarchs **134**, *Pl. 38*
mosquitos 8, 109, 110,
  110-111, *Pl. 24*
  anopheline 8
moths 11, 99, 109, 110,
  114, 123, 124, 137, 139
  clothes **125**
  death's head 128
  emperor **127-128**, *Pl. 32*
  hawk **128**, *Pl. 32*
  jumping bean 126
  mopane 124
  owl **129-130**, *Pl. 33*
  plume **126-127**, *Pl. 31*
  processionary *Pl. 33*
  sphinx **128**, *Pl. 32*
  tiger **130-131**, *Pl. 34*
  wax 126

owl flies **93**, *Pl. 14*

paussids **97**, *Pl. 15*
praying mantids **68-69**, 87,
  89, 92, *Pl. 2*
psocids **79**, *Pl. 5*
psylla, citrus 138
psyllids **84**, *Pl. 8*
pyralids **126**, *Pl. 31*

reds **134-135**, *Pl. 38*

scale insects 82, **85-86**,
  101, 135, 138
  citrus red 138
  pernicious 138
  soft *Pl. 9*
scorpionflies **108-109**, *Pl. 23*
silverfish **65-66**, 66, *Pl. 1*
skippers **131-132**, *Pl. 36*
springtails 65
stick insects **68**, *Pl. 2*
stoneflies **78-79**, *Pl. 5*
stylopids **108**, *Pl. 23*
swallowtails **131**, *Pl. 35*

citrus 131
mocker 131
swifts **133-134**, *Pl. 37*

termites 65, **70-73**, 99, 143,
  Pl. 3
  dry-wood **72**
  harvester **71**
  snouted 73, *Pl. 3*
thrips **90**, *Pl. 12*
tok-tokkies 103
treehoppers **84**, *Pl. 8*

velvet ants **138-139**, *Pl. 40*

wasps 114, 115, 122, 124,
  126, 135, 136
  chalcid **137-138**, *Pl. 9*
  cuckoo **138**, *Pl. 40*
  fig 137
  mason **139**, *Pl. 40*
  potter **139**, *Pl. 40*
  social **139**, *Pl. 41*
  spider-hunting **139**, *Pl. 40*
water striders **88**, *Pl. 11*
water-boatmen **89**, *Pl. 11*
water-scorpions 89, **89**, 90,
  Pl. 11
web-spinners **78**, *Pl. 5*
weevils, seed **106**, *Pl. 22*
  snout **107**, *Pl. 22*
  straight-snouted **108**,
  Pl. 23
whites **132**, *Pl. 36*
woolly bears **130-131**, *Pl. 34*
worms, African army 130,
  Pl. 33
  bag- **124-125**, *Pl. 31*
  boll- 129
  cut- 129
  lesser army 130
  mopane 128
  processionary **129**, *Pl. 33*
  red boll- 130
  tomato 130
  wattle bag- 125

153

INDEX TO SCIENTIFIC NAMES

Numbers in **bold** indicate the main entry, and numbers in *italics* refer to the colour plates.

Acherontia atropos 128
Achroia grisella 126
Acraeidae **134-135**, *Fl. 38*
Acrididae **74-75**, *Pl. 3*
Actorniphilus 80
Aculeata 136
Adephaga 95, **95-98**, 98
Aedes 111
  A. aegypti 111
  A. furcifer 111
Agaonidae 137
Agrotis segetum 129
Aleochara 99
Allodape 141
Alticinae *Pl. 22*
Amblycera **80**, 80
Anachalcos convexus 104
Anaphe reticulata 129, *Pl. 33*
Ancala 114
Anisoptera 67, *Pl. 1*
Anopheles 111, *Pl. 24*
  A. arabiensis 111
  A. funestus 111
  A. gambiae 111
Anophelinae 111
Anoplura **81**, *Pl. 6*
Anthia 96, *Pl. 15*
Aphelinidae 138
Aphididae **85**, *Pl. 8*
Apidae **141-142**, *Pl. 42*
Apis mellifera 141
Apocrita 136, **136-145**
Archaeognatha **65-66**, 66
Arctiidae **130-131**, *Pl. 34*
Arcyophora longivalvis 129
  A. patricula 129
  A. zanderi 129
Argema mimosae 128, *Pl. 32*
Arytaina mopani 84, *Pl. 8*
Ascalaphidae **93**, *Pl. 14*
Asilidae **114**, *Pl. 26*
Atylotus 114
Auchmeromyia bequaerti 119
  A. luteola 119

Avaritia 112

Belenois aurota Pl. 36
Belostomatidae **89-90**, *Pl. 12*
Bembix 140
Bittacidae 108
Blattodea **69-70**, *Pl. 2*
Bombyliidae **114-115**, *Pl. 26*
Bostrychidae **101-102**, *Pl. 19*
Brachycera **113-115**
Brachytrupes membranaceus 76, *Pl.4*
Brentidae **108**, *Pl. 23*
Bromophila caffra 122, *Pl. 30*
Bruchidae **106**, *Pl. 22*
Buprestidae **100-101**, *Pl. 18*
Busseola fusca 130

Cacodmus 88
Cactoblastis cactorum 11, 126, *Pl. 31*
Caelifera **74-75**
Calliphoridae **118-120**, *Pl. 28*
Calpe 129
  C. eustrigata 129
Camponotus 145
Carabidae **96**, 97, *Pl. 15*
Cassidinae 106, *Pl. 22*
Cecidomyiidae **113**, *Pl. 25*
Cerambycidae **105-106**, *Pl. 21*
Ceratina 141
Ceratophaga vastella 125, *Pl. 31*
Ceratopogonidae **112**, *Pl. 24*
Cercopidae **83**, 84, *Pl. 7*
Cetoniinae **104-105**, *Pl. 20*
Chalcidoidea **137-138**, *Pl. 39*
Charaxidae **133-134**, *Pl. 37*
Charaxes brutus Pl. 37
Chironomidae 110, **112**, *Pl. 25*

Chrysididae **138**, *Pl. 40*
Chrysomelidae **106**, *Pl. 22*
Chrysomya 119
  C. albiceps 119, *Pl. 28*
  C. marginalis 119
  C. putoria 119
Chrysopidae **91-92**, 92, *Pl. 13*
Chrysops 114
  C. dimidiata 113
Cicadellidae **83-84**, *Pl. 7*
Cicadidae **82-83**, *Pl. 7*
Cicindelidae **96**, *Pl. 15*
Cimex hemipterus 87
  C. lectularius 87
Cimicidae **87-88**, *Pl. 10*
Cleridae **99-100**, *Pl. 17*
Cobboldia chrysidiformis 117
  C. loxodontis 117
Cobboldiinae 117
Coccidae **85-86**, 133, *Pl. 9*
Coccinellidae **101**, *Pl. 18*
Coleoptera **95-108**, 124, *Pl. 15-23*
Collembola **65**
Colpocephalum 80
Conopidae **122**, *Pl. 30*
Copris 103
Cordylobia anthropophaga 119
Coreidae **86-87**, *Pl. 9*
Corixidae **89**, *Pl. 11*
Corydalidae 91
Crematogaster 144
Ctenocephalides canis 94
  C. felis 94
Ctenolepisma longicaudata 66
Ctenuchidae **130**, *Pl. 34*
Culicidae **110-111**, *Pl. 24*
Culicoides 112
  C. austeni 112
  C. grahamii 112
  C. imicola 112, *Pl. 24*
Curculionidae 95, 106, **107**, 108, *Pl. 22*

154

Cyclorrhapha 99, **115-123**
Cydnidae 86, *Pl. 9*
Cyligramma latona *Pl. 33*
Cyrtacanthacris
 (Nomadacris)
 septemfasciata 74
Cyrtobagus salviniae 107

Damalinia 80
Danaidae 131, **134**, *Pl. 38*
Danaus chrysippus 134,
 Pl. 38
Dendroctonus 107
Dennyus 80
Dermaptera **77-78**, *Pl. 5*
Dermestes maculatus 101
Dermestidae **101**, *Pl. 18*
Diaspididae **85-86**, *Pl. 9*
Diopsidae **122**, *Pl. 30*
Diparopsis castanea 130
Diplura **65**
Diptera 95, 98, 99, 108,
 109-123, *Pl. 24-30*
Ditrysia **124-135**
Dorylinae 144
Dorylus Pl. 43
 D. helvolus 144
Dugari 129
Dynastinae **104**, *Pl. 20*
Dytiscidae **97**, *Pl. 15*

Echidnophaga 94
 E. aethiops 95
 E. gallinacea 94
 E. larina 95, *Pl. 14*
Ectrichodia crux 87
Elateridae **100**, *Pl. 17*
Embiidina **78**, *Pl. 5*
Emporia melanobasis 126
Empusidae *Pl. 2*
Encyrtidae 138
Ensifera **75-77**
Ephemeroptera **66**, 78, *Pl. 1*
Eumenidae **139**, *Pl. 40*

Fannia canicularis 116
Forcipomyia 112
Forficulina 77, 78
Formicidae 70, **142-145**,
 Pl. 43
Formicinae 144
Fulgoridae **83**, *Pl. 7*

Galerucella 106

Galerucinae 106
Galleria mellonella 126
Galleriinae 126
Gasterophilidae **117**, 117,
 Pl. 27
Gasterophilinae 117
Gasterophilus 117, *Pl. 27*
 G. haemorrhoidalis 117
 G. inermis 117
 G. meridionalis 117
 G. nasalis 117
 G. pecorum 117
 G. ternicinctus 117
Gedoelstia 118, *Pl. 27*
 G. cristata 118
 G. haessleri 118
Geometridae **127**, *Pl. 32*
Gerridae **88**, *Pl. 11*
Glossina 121
 G. morsitans 121
 G. pallidipes 121
Glossinidae **120-121**, *Pl. 29*
Gonimbrasia belina 124, 128
Gryllidae **76-77**, *Pl. 4*
Gryllotalpa africana 77
Gryllotalpidae **77**, *Pl. 4*
Gryllus bimaculatus 76
Gymnopleurus 103
Gyna 70
Gyrinidae **97-98**, *Pl. 15*
Gyrostigma 117
 G. pavesii 117

Haematobia spinigera 116
 H. thirouxi potans 116
Haematobosca 116
Haematomyzus elephantis
 80, 81, *Pl. 6*
 H. hopkinsi 80
Haematopinus 81
 H. phacocoeri Pl. 6
Haematopota 114
Hebardina 70
Heliocopris 103
Heliothis armigera 129
Hemerobiidae **91-92**
Hemimerina (Diploglossata)
 77
Hemimerus 78
Hemiptera **82-90**, 144,
 Pl. 7-12
Hesperiidae **131-132**, *Pl. 36*
Heteroptera 82, **86-90**
Hetrodinae 76, *Pl. 4*

Hippoboscidae **121-122**,
 Pl. 29
Hister 98
Histeridae **98**, *Pl. 16*
Hodotermes mossambicus
 71
Hodotermitidae **71**
Hohorstiella 80
Homoptera **82-86**, 86, 133
Hoplistomerus 114
Hymenoptera 70, 95,
 135-145, *Pl. 39-43*
Hymenopodidae *Pl. 2*
Hyperechia 114
Hypoderminae 118

Ichneumonidae **136-137**,
 Pl. 39
Insecta 65, **65-145**
Ips 107
Ischnocera **80**
Isoptera **70-73**, *Pl. 3*

Jashinea 114

Kalotermitidae **72**
Kirkioestrus 118
Kheper 103
Kotochalia junodi *Pl. 14*

Laemobothrion maximum 80
Lagaropsylla idae *Pl. 14*
Lampyridae **99**, *Pl. 16*
Lasiohelea 112
Lauxaniidae **115**, *Pl. 27*
Lepidoptera **124-135**, 135,
 136, *Pl. 31-39*
Leptoconops 112
Leptopsyllus segnis 94
Linognathus 81
Lipoptena paradoxa *Pl. 29*
Liposcelis 79
Locustana pardalina 74, 75
Lucilia cuprina 119
 L. sericata 119
Lycaenidae **132-133**, *Pl. 36*
Lycidae **99**, *Pl. 16*
Lyctidae **101-102**
Lygaeidae **87**, *Pl. 9*
Lymexylidae **100**, *Pl. 17*

Macrotermitinae 72
Macrotermes natalensis
 Pl. 3

Macrotoma Pl. 21
  M. natala 105
Mallophaga Pl. 6
Mansonia Pl. 24
Mantichora 96
Mantispidae **92**, *Pl. 13*
Mantodea **68-69**, *Pl. 2*
Mecoptera **108-109**, *Pl. 23*
Megachilidae **140-141**,
  Pl. 41
Megaloptera **91**, *Pl. 12*
Megaponera foetens 71,
  143, *Pl. 43*
Meloidae **102**, *Pl. 19*
Melophagus ovinus 122
Membracidae **84**, 144, *Pl. 8*
Meromenopon 80
Mesomyia 114
Miridae **88**, *Pl. 10*
Musca 116
  M. domestica 115, 116
  M. domestica calleva 116
  M. sorbens 116
  M. tempestatum 116
Muscidae **115-116**, *Pl. 27*
Muscinae 115
Mutillidae **138-139**, *Pl. 40*
Myrmeciinae 143
Myrmeleontidae **92-93**, 93,
  Pl. 13
Myrmicinae 144
Myrsidea 80

Nasonia vitripennis 138
Nasutitermitinae 73, *Pl. 3*
Necrobia rufipes 100, *Pl. 17*
Nematocera **109-113**, 113
Nemouridae 79
Neochetina eichorniae 107
Neocuterebrinae 117
Neodiplogrammus
  quadrivittatus 107
Neohydronomus pulchellus
  107
Nepidae **89**, *Pl. 11*
Neuroptera 66, 91, **91-93**,
  Pl. 13-14
Noctuidae **129-130**, *Pl. 33*
Nosopsyllus fasciatus 94
Notonectidae **88-89**, *Pl. 11*
Numidicola 80
Nymphalidae **133**, *Pl. 37*

Odonata **66-67**, 91, *Pl. 1*

Oecanthinae 76
Oestridae **117-118**, *Pl. 27*
Oestrinae 118
Oestrus 118, *Pl. 27*
  O. aureoargentatus 118
  O. ovis 118
  O. variolosus 118
Onthophagus 103, 104
Orthoptera (Saltatoria)
  73-77, *Pl. 3-4*
Othreis 129

Papilio demodocus Pl. 35
Papilionidae **131**, *Pl. 35*
Parasitica (Terebrantia) 136
Passeromyia 116
  P. heterochaeta 116
Paussidae **97**, *Pl. 15*
Pediculus capitis 81
  P. humanus 81
Pentatomidae **86**, *Pl. 10*
Pephricus livingstonei Pl. 9
Perlidae 79
Pharyngobolus 118
  P. africanus 118
Phasmatodea **68**, *Pl. 2*
Philoliche 114
Philopteridae 80
Phlebotominae 110
Phlebotomus 110
Phthiraptera **79-81**, *Pl. 6*
Phycitinae 126
Phymateus 75
Pieridae **132**, *Pl. 36*
Piophila casei 123, *Pl. 30*
  P. megastigmata 123
Piophilidae **123**, *Pl. 30*
Platystomatidae **122**, *Pl. 30*
Plecoptera **78-79**, *Pl. 5*
Polyctenidae **87-88**, *Pl. 10*
Polyphaga 95, **98-108**
Polyplax 81
Pompilidae **139**, *Pl. 40*
Ponerinae 143
Princeps dardanus cenea
  131
  P. demodocus 131
Protura **65**
Pseudocreobotra wahlbergi
  Pl. 2
Psocoptera **79**, *Pl. 5*
Psychidae **124-125**, *Pl. 31*
Psychodidae **110**, *Pl. 24*
Psyllidae **84**, *Pl. 8*

Pteromalidae 138
Pteromalus Pl. 39
Pterophoridae **126-127**,
  Pl. 31
Pthirus pubis 81
Pulex irritans 94
Pyralidae **126**, *Pl. 31*
Pyrgomorphidae **75**, *Pl. 3*
Pyrrhocoridae **87**, *Pl. 10*

Reduviidae **87**, *Pl. 10*
Rhinoestrus 118
Rhinomusca dutoiti 116,
  Pl. 27
Rhynchophthirina **80-81**,
  Pl. 6
Rutelinae **105**, *Pl. 20*
Rutteniinae 117

Saprinus 98
Sarcophagidae **120**, *Pl. 28*
Saturniidae **127-128**, *Pl. 32*
Satyridae **135**, *Pl. 38*
Scarabaeidae 95, **103-105**,
  Pl. 19-20
Scarabaeinae (Coprinae)
  103-104, *Pl. 19*
Scarabaeus 103
Schistocerca gregaria 74
Scolytidae **106-107**, *Pl. 22*
Sergentomyia 110
  S. squamipleuris 110
Serodes 129
Sialidae 91
Siphonaptera **93-95**, *Pl. 14*
Sphecidae **140**, *Pl. 41*
Sphecodemyia 114
Sphingidae **128**, *Pl. 32*
Spodoptera exempta 130,
  Pl. 33
  S. exigua 130
  S. littoralis 130
Staphylinidae 77, **98-99**,
  Pl. 16
Stomoxyinae 115
Stomoxys 116
  S. calcitrans 116
Strepsiptera **108**, *Pl. 23*
Strobiloestrus clarkii 118
Stygeromyia maculosa 116
Symphyta **136**
Syrphidae **115**, *Pl. 26*

Tabanidae **113-114**, *Pl. 25*

Tabanocella 114
Tabanus 113, 114
Tachinidae **120**, *Pl. 29*
Tenebrionidae **102-103**,
 Pl. 19
Tenthredinidae **136**, *Pl. 39*
Termitidae 70, **72-73**, *Pl. 3*
Termitinae 72
Tetrasphondylia terminaliae
 113, *Pl. 25*
Tettigoniidae **76**, *Pl. 4*
Thaumetopoeidae **129**,
 Pl. 33
Thriambeutes 114
Thysanoptera **90**, *Pl. 12*
Thysanura **65-66**, 66, *Pl. 1*
Tinea pellionella 125

Tineidae **125**, *Pl. 31*
Tineola bisselliella 125
Tipulidae 108, **109-110**,
 Pl. 24
Trichapion lativenne 107
Trichodectes 80
Trichodectidae 80
Trichophaga tapetziella 125
Trichoptera **123-124**, *Pl. 30*
Trigona 142
Trogidae **105**, *Pl. 20*
Trox melancholicus 105
 T. mutabilis 105
 T. radula 105
 T. rusticus 105
 T. squalidus 105
 T. tuberosus 105

Tunga penetrans 95

Utetheisa pulchella Pl. 34

Vespidae **139-140**, *Pl. 41*

Xenopsylla 8, 94
Xylocopa 114, 141
Xylocopidae **141**, *Pl. 42*

Zonocerus elegans 75,
 Pl. 3
Zoraptera **73**
Zutulba zelleri Pl. 31
Zygaenidae **125-126**, 130,
 Pl. 31
Zygoptera 67, *Pl. 1*

INDEX TO AFRIKAANS NAMES

Numbers in **bold** indicate the main entry.

bergstroomjuffers **91**
besies **82-90**
　grysstink- **87**
　lantern- **83**
　skildstink- **86**
　skuim- **83**
　son- **82-83**
blaarspringers **83-84**
blaartrektorre **102**
blaaspootjies **90**
bloutjies **132-133**
boktorre **105-106**
boomspringers **84**
bootmannetjies **89**
brommers **118-120**
bye 135
　blaarsny- **140-141**
　heuning- **141-142**
　hout- **141**
bytende muggies **112**

dubbelsterte **133-134**

galmuggies **113**
glimwurms **99**

hotnotsgotte **68-69**

kakkerlakke **69-70**
kewers **95-108**
　bas- **106-107**
　blaarvreet **106**
　bontroof- **99-100**
　boontjie- **106**
　dwaal- **98-99**
　grond- **96**
　harlekyn- **98**
　houtpoeier- **101-102**
　langhoring- **105-106**
　platvlerk- **99**
　prag- **100-101**
　renoster- **104**
　sand- **96**
　skemer- **102-103**
　snuit- **107**
　valssnuit- **108**
　vel- **101**
　water- **97**
kniptorre **100**

kokerjuffers **123-124**
kopertjies **132-133**
krenkelvlerke **108**
krieke 73
　koring- **76**
　mol- **77**
　veld- **76-77**

landmeters **127**
langpote **109**
luise **79-81**
　boek- **79**
　dop- **85-86**
　plant- **85**
　wee- **87**

miere 135, **142-145**
　fluweel- **138-139**
mierleeus **92-93**
　langhoring- **93**
miskruiers **103-104**
motte 124
　klere- **125**
　tier- **130-131**
　uil- **129-130**
　veer **126-127**
　wol- **125**
motvliegies **110**
muskietmuggies **112**
muskiete **110-111**

naaldekokers **66-67**
netvlerkies **91-92**
netvlerkinsekte **91-93**

oorkruipers **77-78**
perdebye 135
　pleister- **139**
pouoë **127-128**
pylsterte **128**

rondekophoutboorders **101-102**
rooitjies **134-135**
rugswemmers **88**

sakwurms 124
silwervissies **65-66**
skilpadjies **101**

skoenlappers 124
　borselpoot- **133**
　melkbos- **134**
spinnekopjagters **139**
sprinkane 73
　korthoring- **74**
　langhoring- **76**
stokinsekte **68**
swaelsterte **131**

termiete **70-73**
　droëhout- **72**
　hodograsdraer- **71**
tonnelspinners **78**

valshotnotsgotte **92**
vismotte **65-66**
vlieë **109-123**
　blinde- **113**
　by- **114**
　dikkop- **122**
　eendags- **66**
　huis- **115-116**
　luis- **121-122**
　pêrel- **78-79**
　roof- **114**
　rooikop- **122**
　skerpioen- **108**
　steeloog- **122**
　sweef- **115**
　tsetse- **120-121**
　vleis- **120**
vlooie **93-95**

wantse, blaarpoot-
　86-87
　graaf- **86**
　plant- **88**
　reusewater- **89**
　roof- **87**
　rooi- **87**
　sluipmoord- **87**
waterhondjies **97-98**
waterjuffers **66-67**
waterlopers **88**
waterskerpioene **89**
wespe, blad- **136**
　koekoek- **138**
witjies **132**